D0918854

LANDFILL BIOREACTOR DESIGN AND OPERATION

Debra R. Reinhart
Timothy G. Townsend

Landfill Bioreactor Design and Operation

LEWIS PUBLISHERS

A CRC Press Company
Boca Raton   London   New York   Washington, D.C.

Library of Congress Cataloging-in-Publication Data

Reinhart, Debra R.
    Landfill Bioreactor design and operation / Debra R. Reinhart, Timothy G. Townsend
        p. cm.
    Includes bibliographical references and index.
    ISBN 1-56670-259-3
    1. Bioreactor landfills. I. Townsend, Timothy G. II. Title
TD795.7.R45 1997
628.4′4564—dc21   97-22220

Visit the CRC Press Web site at www.crcpress.com

© 1998 by CRC Press LLC
Lewis Publishers is an imprint of CRC Press LLC

No claim to original U.S. Government works
International Standard Book Number 1-56670-259-3
Library of Congress Card Number 97-22220
Printed in the United States of America   4 5 6 7 8 9 0
Printed on acid-free paper

THE AUTHORS

 **Dr. Debra R. Reinhart** is an Associate Professor in the Department of Civil and Environmental Engineering of the University of Central Florida (UCF). In 1996, she became Associate Dean for Research for the College of Engineering. Dr. Reinhart received a B.S. in Environmental Engineering from UCF and M.S. and Ph.D. degrees in Environmental Engineering from the Georgia Institute of Technology.

Before joining the University of Central Florida faculty, Dr. Reinhart spent five years in the consulting engineering field serving as a process engineer. She also served for five years with the city of Atlanta's Research and Development Division of the Bureau of Pollution Control.

Dr. Reinhart's research interests include solid waste management and groundwater remediation. Currently she is conducting research on leachate recirculation, liner clogging, leachate depth on liner, and waste composition analysis. She also has investigated gas emission measurement and reuse, and yard waste composting. In the groundwater remediation field, she is investigating chlorinated solvent treatment using reactive walls and electrokinetic soil treatment for metals removal. She holds patents in both fields.

Dr. Reinhart is an active member of the Solid Waste Association of North America, Water Environment Federation, Air and Waste Management Association, Association of Environmental Engineering Professors, the International Association of Water Quality, and American Society of Civil Engineers. She served on the boards of directors of the Florida Center for Solid and Hazardous Waste Management and the Florida Air and Waste Management Association. She is a registered professional engineer in Florida and Georgia and a Diplomate of the American Academy of Environmental Engineers. Dr. Reinhart has authored more than 30 publications in journals and proceedings.

Timothy G. Townsend is an Assistant Professor in the Department of Environmental Engineering Sciences at the University of Florida. Dr. Townsend received a Ph.D. from the University of Florida in 1996, after spending six years conducting research at the Alachua County Southwest Landfill on bioreactor landfill operation and engineering.

He currently conducts research in all areas of solid waste management and engineering, and teaches a number of solid and hazardous waste related management and design courses.

PREFACE

Research into the use of the municipal solid waste landfill as a bioreactor has been conducted for more than 20 years. The research has progressed from laboratory-scale reactors to pilot-scale lysimeters to recent full-scale demonstration of landfill bioreactor technology. Studies have shown that bioreactor operation provides an opportunity to control waste decomposition within the landfill environment, minimizing long-term risk to human health and the environment.

Because of the importance of encouraging rapid waste stabilization within a landfill, the bioreactor landfill is central to many waste management plans today. The implementation of bioreactor technology at full-scale landfills to date, however, has been accomplished through application of information gathered anecdotally, to the great frustration of designers and operators. The need for organized and complete data regarding the bioreactor landfill clearly was expressed at a gathering of more than 200 professionals at a bioreactor landfill seminar sponsored by the U.S. Environmental Protection Agency in 1995. This book is the outcome of many years of landfill bioreactor research by the authors both in the laboratory and in the field. It is the result of long discussions with professionals studying, designing, and operating the landfill as a bioreactor. This book is offered as a summary of design and operating experiences of professionals all over the world engaged in the safe disposal of municipal solid waste, while attempting to avoid perpetual storage of that waste.

This book is not a design guide for the complete sanitary landfill. There are many fine and detailed texts available that should be consulted to complement this book. However, Chapter 2 does provide some fundamentals concepts offered to provide a frame of reference to the reader. Some important features of the book include a detailed description of laboratory, pilot, and full-scale bioreactor studies, case studies of full-scale operating landfills with pictures and schematics to aid in design, and a discussion of the current understanding of moisture routing through the landfill. The value of past research efforts is recognized in each and every chapter as the authors attempt to provide design and operational insight gleaned from these studies. Finally, the text provides strategy to design a sustainable waste management plan through recovery of stabilized waste and reuse of the landfill site.

ACKNOWLEDGMENTS

Much of the information in this text was gathered and adapted from numerous other waste professionals. Unfortunately, there are too many to name all of them here.

We would like to acknowledgment the contributions of a few, however, including Phillip McCreanor, Santosh Agora, Mark James, Basel Al-Yousfi, and David Carson. The guidance from Drs. Fred Pohl and Lamar Miller could never be forgotten. And, of course, we must thank our spouses, Richard Reinhart and Marlene Townsend, for their patience and tolerance over the years.

CONTENTS

INTRODUCTION

1

■ SCOPE AND OBJECTIVES

While the typical person may think otherwise, the modern municipal solid waste (MSW) landfill has evolved into a sophisticated facility. The state-of-the-art landfill can be loosely categorized into four classes. The *secure landfill* tends to entomb wastes, perhaps postponing any environmental impact to the future when environmental controls and safeguards initially provided fail. The *monofill* accepts waste that cannot be processed otherwise through resource recovery, composting, or incineration. These materials tend to be inert and may be more easily assimilated by the environment. At present, the monofill is used for disposal of combustion ash, construction and demolition debris, and yard waste. The *reusable landfill* permits excavation of the landfill contents to recover metals, glass, plastics, other combustibles, compost and, potentially, the site itself following a lengthy stabilization period. A fourth, emerging landfill type, is the *bioreactor landfill*, which is operated in a manner to minimize environmental impact while optimizing waste degradation processes. It is this type of landfill that is the focus of this book.

Bioreactor landfills are constructed in a manner similar to most modern sanitary landfills — they are equipped with liners and leachate collection systems. Bioreactor landfills, however, are operated and controlled to rapidly accelerate the biological stabilization of the landfilled waste. Fig. 1.1 provides a schematic of a typical bioreactor landfill operation. The fundamental process used for waste treatment in a bioreactor landfill is leachate recirculation. Recirculation, or recycle, of leachate back to the landfill creates the environment favorable to rapid microbial decomposition of the biodegradable solid waste. The landfill becomes a treatment system rather than simply a storage facility. Bioreactor operation thus provides additional protection of the environment and may reduce long-term liability and associated monitoring costs. The rapid treatment of the waste also facilitates the operation of a bioreactor landfill as a *reusable landfill*. The combined operation of rapid waste treatment followed by reclamation of the stabilized landfill material results in a system that may dramatically extend the operating life beyond that of a traditional landfill.

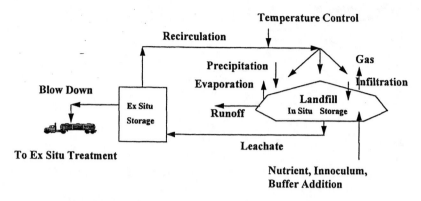

Figure 1.1 Schematic diagram of the landfill bioreactor.

Laboratory and pilot-scale studies have clearly demonstrated that operation of a landfill as a bioreactor accelerates waste degradation, provides *in situ* treatment of leachate, enhances gas production rates, and promotes rapid settlement. Full-scale evidence of bioreactor landfill benefits is rapidly accumulating. However, many challenges still face the implementation of bioreactor technology, including regulator reluctance to permit such facilities, the availability of theoretical design criteria, the ability to uniformly wet the waste, and operator training issues. The objective is to provide the regulator, designer, landfill owner, and operator with information and data that support the utility of the landfill bioreactor and provide design and operating criteria essential for the successful application of this technology.

◼ THE EVOLUTION OF LANDFILLS FOR WASTE MANAGEMENT

The landfill, as we know it, has evolved from a long tradition of land disposal of MSW, dating back to prehistoric times. Problems with land disposal began as society developed and population density increased. Land disposal of waste — often as open dumps — was subject to aesthetic, safety, and health problems that prompted innovations in design and operation. Environmental impacts associated with MSW landfills have complicated siting, construction, and operation of the modern landfill. Production of leachate has lead to documented cases of groundwater and surface water pollution. Landfill gas emissions can lead to malodorous circumstances, adverse health effects, explosive conditions, and global warming. Traffic, dust, animal and insect vectors of disease, and noise often are objectionable to nearby neighbors. These issues have lead to strict regulation of landfill disposal of MSW.

Ideally, land should be a repository exclusively for inert "earthlike" materials that can be assimilated without adverse environmental impact, a conviction held by landfill regulators, designers, and operators through-out the world.[1] However, successful application of this concept requires extensive waste processing to develop an acceptable product for land disposal, and faces challenges related to public opposition, economics, waste handling, and transportation of recovered materials. The most reasonable scenario for success appears to be a MSW management park, where regional facilities for managing waste — from separation to resource recovery/reuse to incineration to landfilling — would be collec-tively sited.

While disposal solely of inert materials may be an admirable objective, it will be some time, if ever, before this concept can be universally applied. Therefore, it is likely that landfills will continue to receive a variety of materials with potential for environmental impact. A second global consensus is that where leachable materials are land disposed, impenetrable barriers must be provided and waste stabilization must be enhanced and accelerated so as to occur within the life of these barriers. That is, the landfill must be designed and operated as a bioreactor. Additional advantages of the bioreactor landfill include increased gas production rates over a shorter duration, improved leachate quality, and more rapid landfill settlement.

■ LANDFILLS AS BIOREACTORS

Under proper conditions, the rate of MSW biodegradation can be stimu-lated and enhanced. Environmental conditions that most significantly impact biodegradation include pH, temperature, nutrients, absence of toxins, moisture content, particle size, and oxidation-reduction potential. One of the most critical parameters affecting MSW biodegradation has been found to be moisture content. Moisture content can be most prac-tically controlled via leachate recirculation. Leachate recirculation provides a means of optimizing environmental conditions within the landfill to provide enhanced stabilization of landfill contents as well as treatment of moisture moving through the fill. The numerous advantages of leachate recirculation include distribution of nutrients and enzymes, pH buffering, dilution of inhibitory compounds, recycling and distribution of methano-gens, liquid storage, and evaporation opportunities at low additional construction and operating cost. It has been suggested that leachate recirculation can reduce the time required for landfill stabilization from several decades to two to three years.[2]

Implementation of bioreactor technology requires modification of design and operational criteria normally associated with traditional land-filling. For example, the bottom liner system must be designed to accom-modate the additional flows contributed by leachate recirculation. The gas management facilities must be operated to control amplified gas production, particularly during the active landfill phase. Overcompaction of waste during placement may adversely impact leachate routing and prevent even moisture distribution. Use of permeable alternative daily cover is recommended for similar reasons. Leachate recirculation devices must be employed which are compatible with daily operation and closure requirements. Monitoring of leachate and gas quality and quantity becomes critical to operational decision making. Even waste pretreatment, such as shredding or screening, may be desirable to promote biological and chemical landfill processes. Leachate management must be carefully planned to ensure adequate supply for recirculation during dry weather conditions, and on the other extreme, to prevent saturation of waste during wet weather periods. Chronological changes in leachate characteristics will impact *ex situ* treatment and disposal requirements as a result of *in situ* treatment of degradable organics, as well as many hazardous leachate constituents.

■ REGULATORY STATUS

Regulations promulgated under Subtitle D of the Resources Conservation and Recovery Act (RCRA) allow leachate recirculation, provided a composite liner and leachate collection system are included in the design. In the preamble to Subtitle D regulations, implemented in 1991,[5] EPA commented that:

> ... EPA recognizes that landfills are, in effect, biological systems that require moisture for decomposition to occur and that this moisture promotes decomposition of the wastes and stabilization of the landfill. Therefore, adding liquids may promote stabilization of the unit...

Specifically, Section 258.28(b)(2) of CFR Part 258, "Criteria for Municipal Solid Waste Landfills" states the following:

> Bulk or noncontainerized liquid waste may not be placed in a municipal solid waste landfill unit unless ... the waste is leachate or gas condensate derived from the municipal solid waste landfill unit and the landfill unit is equipped with a composite liner and a leachate collection system that is designed and constructed to maintain less than a 30-cm depth of leachate over the liner.

A telephone poll of U.S. state regulatory agencies conducted in late 1992 and early 1993 found that full-scale operation of leachate recirculation was practiced (or would be soon) in 12 states and was permissible in all but seven states. Recirculation facilities were in place at landfills in eight states, under construction at landfills in four states, and planned at landfills in several other states. For the most part, states merely adopted RCRA criteria regarding leachate recirculation, requiring a composite liner in order to implement leachate recirculation.

A few states identified additional, more stringent criteria for leachate recirculation. For example, Florida, Georgia, Pennsylvania, and Virginia specifically address odor prevention. Florida requires that gas management facilities be in place prior to commencement of leachate recirculation. New York requires a double composite liner for all MSW landfills. Pennsylvania and Georgia require that the leachate recirculation piping system be installed under a permeable intermediate cover. Virginia, New York, Georgia, and Florida require control of runoff and prohibit ponding. Georgia also requires that sufficient waste be in place to provide sufficient moisture absorption prior to initiating recirculation.

Those states that prohibited leachate recirculation did so for several reasons. Regulators cited a lack of confidence in the method, interference with the leachate collection system, geological and climate concerns, freezing problems, leachate seepage, lack of waste absorptive capacity, and exacerbated gas and odor production.

■ ORGANIZATION OF THE BOOK

The long term acceptance of bioreactor landfills as tools in integrated solid waste management systems requires that such facilities be designed and operated in a safe and effective manner. Numerous studies have been performed investigating this technology, and this text serves to bring this information together, so that future bioreactor systems may benefit from the lessons gained from previous work. This text summarizes existing information available regarding the design and operation of bioreactor landfills and should serve as a resource to engineers, regulators, and all parties interested in the technology. The content and organization of the book are indicated in Fig. 1.2. In addition to a review of modern sanitary landfill fundamentals, sections cover the results of previous bioreactor landfill operations experiments, including a series of case studies; describe design and operation issues; and discuss the potential of landfill mining as a method to recover treated waste materials and to reuse the bioreactor cells.

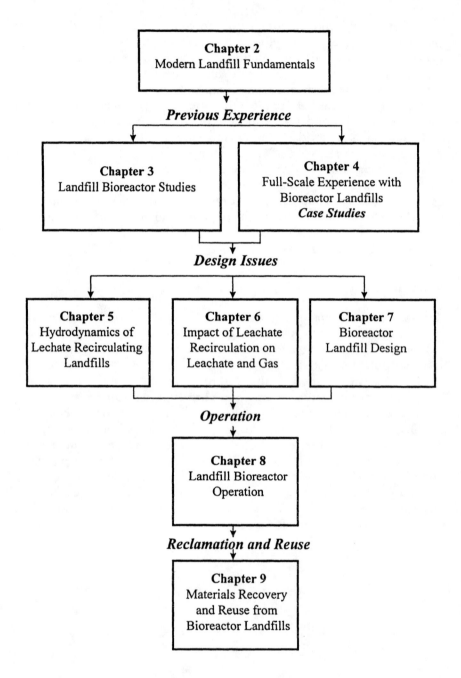

Figure 1.2 Organization of book.

2 MODERN LANDFILL FUNDAMENTALS

INTRODUCTION

A discussion of bioreactor landfills and the use of leachate recirculation as a method of leachate management must begin with an overview of the fundamentals of modern landfill design and operation. Most leachate recirculation operations performed to date have been conducted at traditionally designed landfills. In the future, successful leachate recirculation systems will be engineered as part of an integrated bioreactor landfill design. This chapter provides the reader with a background in the fundamental elements involved in the design of a modern sanitary landfill.

The design and operation of a modern landfill involves many science and engineering disciplines, including the biology and chemistry of waste decomposition and leachate production, as well as the hydraulic, geotechnical, and materials engineering required for the design of liners, pipes, and pumps. While an introduction is provided here, it is beyond the scope of this book to present complete information on every landfill component. A number of texts and research articles are available to provide the reader with detailed information.[4-8] This chapter first will present an overview of the fundamental features of modern sanitary landfills with a review of the containment and control technologies common to all landfills. The unique microbiology and chemistry of MSW landfills and the impact of these characteristics on the design and operation of landfills will then be discussed.

OVERVIEW OF MODERN SANITARY LANDFILLS

The sanitary landfill evolved in response to hazards associated with the indiscriminate dumping of wastes. The health, safety, and aesthetic problems encountered at open dumps of the past included the presence of rodents, flies, fires, and odors. Sanitary landfills developed when controlled operation and disposal techniques such as daily cover and compaction were found to minimize many of the safety and aesthetic concerns. These measures were not sufficient, however, to control other problems associated with leachate and gas production — less obvious but potentially greater threats to human health and the environment. These problems have resulted in the promulgation of strict regulations for the design and operation of new landfills. A conceptual layout of a modern sanitary landfill is presented in Fig. 2.1.

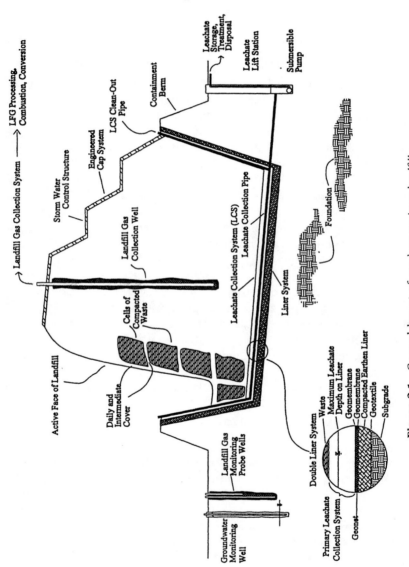

Figure 2.1   General layout of modern sanitary landfill.

Leachate is generated in a landfill as a consequence of the contact of water with solid waste. Leachate may contain dissolved or suspended material associated with wastes disposed of in the landfill, as well as many byproducts of chemical and biological reactions. Leachate from municipal solid waste landfills varies in strength as a result of the biological activity occurring as the solid waste degrades. Leachate from young landfills has both a high dissolved solids content as well as a high concentration of organic matter relative to domestic wastewater. Leachate also may contain a trace amount of hazardous constituents found in the waste stream. In the past, leachate tended to migrate from the landfilled waste, resulting in contamination of the underlying soil and groundwater. The threat of migration led to the requirement, in many countries, of barrier layers to restrict the escape of leachate from the landfill, and to minimize the amount of water that could enter the landfill to create leachate. Once collected, leachate must be removed from the landfill, stored, treated, and disposed.

Landfill gas results from biological decomposition of organic material in the solid waste stream. Approximately 50 to 70 percent of MSW is composed of biodegradable materials. In a landfill, microorganisms act to degrade this material. For most of the landfill life anaerobic conditions dominate, and the primary end products of anaerobic waste decomposition are the gases methane and carbon dioxide. These two gases, along with minor amounts of water and trace components, are termed landfill gas (LFG). LFG creates a number of problems because methane is explosive, many of the trace components have toxic and odor-causing characteristics, and methane serves as a greenhouse gas. On the other hand, landfill gas presents a resource recovery opportunity because of the energy value of methane. As a result, landfill gas must be managed properly. Control of landfill gas is even more critical in environments where gas production is accelerated, i.e., with leachate recirculation.

The necessity to control leachate and landfill gas has resulted in the greatest changes to modern sanitary landfill design and operation. The migration of both leachate and landfill gas to the environment must be minimized. The remainder of Chapter 2 is dedicated to presenting the methods of leachate and LFG control and management at modern sanitary landfills. Topics covered include the containment systems necessary to protect the environment; collection, control, and management of leachate and LFG; and operation strategies for sanitary landfills.

■ LANDFILL CONTAINMENT SYSTEMS

The concept of containment systems for modern sanitary landfills involves the use of barrier layers to prevent leachate from leaving the landfill and contaminating the underlying soil and groundwater, and to prevent water from entering the landfill to create leachate. Barrier layers are constructed of materials that possess a low permeability to water. The most common materials include compacted soil (clay) and synthetic membranes (geomembranes). The containment layer at the bottom of a landfill is known as a liner and the one at the top is referred to as a cap. While conceptually the barrier layers may be thought of as one unit, they are in reality multiple layers of different materials, and are more accurately referred to as liner and cap systems.

Compacted Soil Barrier Layers

Many soils naturally possess characteristics that make them relatively impermeable to water flow. Clayey soils are a good example of a naturally impermeable material. Because of the small particle size and the surface chemistry of clay minerals, clay deposits in the environment greatly restrict the rate of water movement. Natural clay deposits sometimes are used as landfill barrier layers. In most sanitary landfills, however, clay liners are constructed by modifying the structure of the clay soil brought to the site by the addition of water and mechanical compaction to achieve optimum engineering characteristics.

A number of properties make compacted soil amenable to use as a component in a landfill containment system. These include mechanical properties such as shear strength, but most importantly, the permeability of the clay to water. The engineering parameter relating the permeability of a porous media to the flow of water is the hydraulic conductivity. Most engineered clay liners must meet regulatory requirements for hydraulic conductivity of less than 10^{-7} cm/sec. The hydraulic conductivity of a compacted soil, along with many other parameters, must be tested routinely during soil liner construction.

Synthetic Barrier Layers

In recent years an entire field of engineered materials known as geosynthetics has developed. These materials are generally composed of plastics, and are used for a wide variety of waste containment applications. One of the most common uses of geosynthetic material at landfills is the geomembrane. A geomembrane is a thin sheet of plastic that possesses the characteristic of being highly impermeable to water and resistant to chemical attack from the waste it is designed to contain. Long strips of geomembrane material may be attached to one another to form a continuous barrier layer. The most common type of geomembrane used at a landfill is high density polyethylene (HDPE), but other materials in use include polyvinyl chloride (PVC), low density polyethylene (LDPE), and polypropylene (PP). Geomembranes that exhibit the greatest chemical resistance and the lowest water vapor transmission rates (such as HDPE) are normally used as components of bottom liners. Geomembranes that possess a large modulus of elasticity are typically used for components of cap systems.

Geomembranes hold some advantages over clay because they are less permeable to water transport, much less material is required, and they are easy to install. Disadvantages include the fact that they are more susceptible to leaks from damage during installation and their long-term performance is uncertain. For these reasons, many barrier layer applications use combinations of both a geomembrane and a compacted soil liner.

Liner Systems

The use of multiple components in the design of a low permeability barrier layer results in the creation of a liner system. As previously stated, the use of a compacted soil layer and a geomembrane provides an effective combination. Such a liner is defined as a composite liner. Current U.S. federal regulations for MSW landfills (brought about by the Resource Conservation and Recovery Act, RCRA) require a liner system composed of a composite liner with 60 cm (2 ft) of compacted soil at a maximum hydraulic conductivity of 10^{-7} cm/sec and a geomembrane.[3] The geomembrane must be at least 30 mil in thickness (60 mil in the case of HDPE) and the geomembrane must be in intimate contact with the compacted soil. The composite liner then is overlain by a drainage layer that limits the depth of leachate on the liner to less than 30 cm (1 ft) at all times. In addition to barrier layers, however, other layers must be provided to permit the drainage and removal of leachate away from the liner. These leachate collection systems (LCS) are composed of highly permeable materials, either naturally occurring such as a sand or gravel, or synthetic, such as a geonet and geotextile. The design of leachate collection systems will be discussed in a future section. A cross-section of a RCRA Subtitle D liner (MSW landfill) is presented in Fig. 2.2.

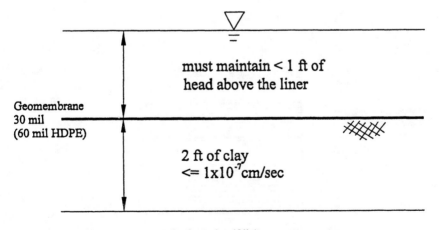

Figure 2.2   Subtitle D landfill liner cross section.

An alternative to the composite liner system is the double liner system. A double liner consists of two barrier layers, with a drainage layer placed between. The upper and lower barrier layers may be either single or composite liners. The most common application is the use of a composite liner as the lower component and a geomembrane or geocomposite (a thin layer of low permeability clay attached to a geomembrane) as the upper component. This configuration is the liner design for a RCRA Subtitle C hazardous waste landfill. The drainage layer above the upper barrier layer is the primary leachate collection system. The drainage layer in between the two barrier layers is the secondary leachate collection system or leak detection system. It is becoming more commonplace for state regulators or landfill design engineers to choose double liner systems for MSW landfills as well.[9]

Cap Systems

A cap system functions in a similar manner as a liner system, except the purpose is to keep water from entering the landfill. The RCRA Subtitle D MSW landfill regulations require landfill closure when the permitted capacity of a landfill is reached. The closure process requires, among other things, that a cap system be installed to prevent entrance of water into the landfill, as well as gas migration. While no technical standards have been issued for the specific components of the cap system, the barrier layer in the cap system must not have a hydraulic conductivity greater than the bottom component of the liner system. A suggested cap design by the U.S. EPA is presented in Fig. 2.3. Drainage layers are also included as part of the cap system to serve as gas venting layers to facilitate gas transport to collection wells. A vegetative layer is located above the barrier layer to prevent surface soil erosion.

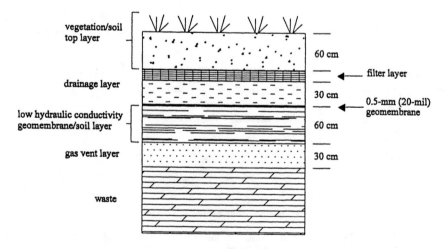

Figure 2.3   EPA recommended landfill cap system.

■ COLLECTION AND CONTROL OF LEACHATE

As previously stated, leachate results from the contact of water with solid waste. While some leachate results from moisture present in the waste when it is disposed, most leachate is generated as a result of rainfall infiltrating into the solid waste. Liner systems are designed and constructed to prevent leachate from migrating into the soil and groundwater underneath the landfill, but this leachate must be properly removed, stored, treated, and disposed. The design of a modern landfill facility requires a prediction of maximum and average leachate generation volumes. A drainage and piping system must be designed to route leachate from the liner and to remove it from the landfill.

Leachate Generation

When rainfall falls on a landfill site, the resulting water either will be shed from the fill as stormwater runoff, evaporate, transpire under the action of surface vegetation, or infiltrate into the landfill to create leachate. Leachate that infiltrates into the landfill either is stored — absorbed — by the landfilled material or it migrates through the landfill under the force of gravity, ultimately being intercepted by the liner system. The amount of leachate generated at a landfill depends on many conditions, including site climate, landfill morphology, waste depth, landfill surface conditions, and the operation of the facility. The prediction of the amount of leachate that is produced at a given landfill is generally estimated by performing a water budget analysis. A simplified water budget for a landfill is presented in Fig. 2.4.

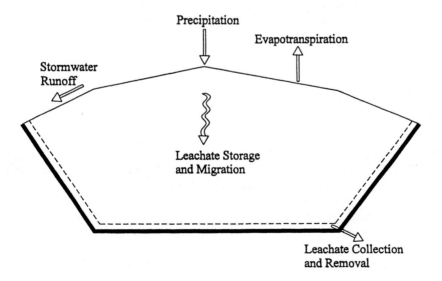

Figure 2.4 Landfill water balance.

A water budget analysis is a common procedure in the field of hydrology. Fenn[10] applied this technique to landfills to predict leachate generation. Standard hydrologic tools are used to determine the amount of rainfall that infiltrates into the landfill for a given set of climate and site conditions. This water at first is stored in the landfilled material. The amount of water which a permeable material such as waste may store against the force of gravity before it drains is defined as field capacity. The available storage, therefore, is the difference between the field capacity and the initial moisture content of the material. In the Fenn water budget model, leachate production begins when field capacity is exceeded for the entire depth of the landfill. At this point the amount of leachate produced is equal to the amount of water infiltrating into the landfill.

A number of more complex landfill water budget models have been developed in recent years.[11-13] These models incorporate more advanced hydrologic techniques and incorporate climatological databases. The simulation of water flow through the landfill to more accurately predict the unsaturated flow conditions which typically occur is a common feature of most modeling routines. The most commonly applied landfill water budget model is the Hydrologic Evaluation of Landfill Performance model.[11] Currently supporting its third version, the HELP model is a quasi-two-dimensional hydrologic model of water movement across, into, through, and out of landfills.

The landfill water budget may be modified to include an input representing leachate recycle. The current version of HELP permits this feature. The hydrologic modeling of leachate recycle systems will be covered in more detail in Chapter 5.

■ LEACHATE COLLECTION AND STORAGE

When leachate is intercepted by the barrier layer of a liner system, it must be routed from the liner through the use of drainage layers. Leakage through a barrier liner, either as a result of permeation through the material or leakage through a hole or imperfection, increases with an increasing depth of leachate on the liner. Leachate collection systems are therefore designed to minimize the depth of leachate (often referred to as head) above the liner. The RCRA Subtitle D landfill regulations limit the depth of leachate to no more than 30 cm (1 ft) at any given time.[3]

The four parameters that have the greatest impact on head above the liner are: flow rate of leachate into the LCS, the permeability of the drainage layer, the length of the drainage path, and the slope of the liner. These parameters are identified in Fig. 2.5 for a typical LCS design. The design engineer uses a model such as HELP to predict leachate flow into the LCS. The materials that make up the LCS are typically permeable soils such as sand and gravel, or synthetic drainage devices such as a geonet.

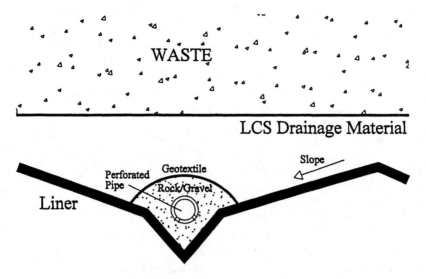

Figure 2.5 Typical leachate collection system profile.

The design engineer than adjusts the spacing of the pipes (drainage path) and the slope of the liner to maintain the leachate level within the regulatory requirement. A number of equations have been developed to predict head on the liner as a function of the variables discussed.[14-15] Such a routine is incorporated into the HELP model.[11]

Leachate drains from the LCS to a series of trenches that typically contain large diameter pipes surrounded by a blanket of gravel. If necessary, a geotextile fabric filter is used to separate the gravel from other drainage materials in the LCS such as sand. The trenches themselves are sloped, and ultimately drain to a sump or lift-station. The sump contains a pump that is used to remove leachate from the landfill. A storage system must be provided for leachate at the landfill site, and may be part of a leachate treatment system. Typical storage systems include lined surface impoundments, and both closed and open tanks.

■ LEACHATE AND LANDFILL GAS MANAGEMENT AT MSW LANDFILLS

The elements discussed regarding landfill containment and leachate generation, collection, and storage, are common to all modern landfill designs. MSW landfills possess characteristics distinct from other types of landfills as a result of the large amount of biodegradable materials that is present in the waste, and the resulting decomposition or stabilization of these materials. The extent of biological activity that occurs in a landfill controls to a large degree leachate quality and therefore dictates the types of treatment processes used. The decomposition of waste also results in the production of biogas as an end product, which in turn must be controlled and may result in an energy recovery opportunity. This section first describes the landfill as a biologically active system, and then explores the unique characteristics and control of the leachate and landfill gas which result from this environment.

The Landfill as a Biological System

Numerous landfill investigation studies have suggested that the stabilization of waste proceeds in five sequential and distinct phases.[16] The rate and characteristics of leachate produced and biogas generated from a landfill vary from one phase to another and reflect the processes taking place inside the landfill. These changes are depicted graphically in Fig. 2.6, and are clearly observed in laboratory-scale simulated landfills. In a large-scale landfill where waste is placed over a lengthy period of time, the waste stabilization phases tend to overlap and the leachate and gas characteristics reflect this phenomenon.

The initial adjustment phase (Phase I) is associated with initial placement of solid waste and accumulation of moisture within landfills. An acclimation period — or initial lag time — is observed until sufficient moisture develops and supports an active microbial community. Preliminary changes in environmental components occur in order to create favorable conditions for biochemical decomposition.

In the transition phase (Phase II), the field capacity is often exceeded, and a transformation from an aerobic to anaerobic environment occurs, as evidenced by the depletion of oxygen trapped within the landfill media. A trend toward reducing conditions is established in accordance with shifting of electron acceptors from oxygen to nitrates and sulfates, and the displacement of oxygen by carbon dioxide. By the end of this phase, measurable concentrations of chemical oxygen demand (COD) and volatile organic acids (VOA) can be detected in the leachate.

In the acid formation phase (Phase III), the continuous hydrolysis (solubilization) of solid waste, followed by (or concomitant with) the microbial conversion of biodegradable organic content, results in the production of intermediate VOAs at high concentrations. A decrease in pH values is often observed, accompanied by metal species mobilization. Viable biomass growth associated with the acid formers (acidogenic bacteria), and rapid consumption of substrate and nutrients are the predominant features of this phase.

During the methane fermentation phase (Phase IV), intermediate acids are consumed by methane-forming consortia (methanogenic bacteria) and converted into methane and carbon dioxide. Sulfate is reduced to sulfide. The pH value is elevated, being controlled by the bicarbonate buffering system, and consequently supports the growth of methanogenic bacteria. Heavy metals are removed from the leachate by complexation and precipitation and transported to the solid phase.

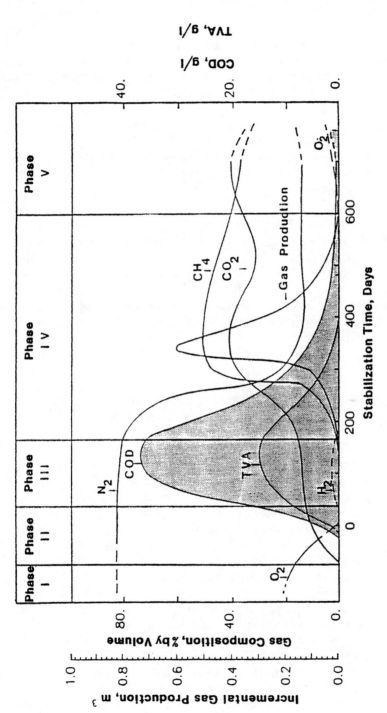

Figure 2.6 Five phases of landfill stabilization. (Adapted from Pohland and Harper, 1986.)

During maturation phase (Phase V), the final state of landfill stabilization, nutrients and available substrate become limiting, and the biological activity shifts to relative dormancy. Gas production dramatically drops and leachate strength remains constant and at much lower concentrations than earlier phases. Because gas production has all but ceased, atmospheric gases may permeate back into the landfill, and oxidized species may slowly appear. The slow degradation of resistant organic fractions may continue with the production of humic-like substances.

Thus, the progress toward final stabilization of landfill solid waste is subject to the physical, chemical, and biological factors within the landfill environment, the age and characteristics of landfilled waste, the operational and management controls applied, as well as the site-specific external conditions.

Characteristics of Leachate

The characterization of leachate provides important information necessary for the control of landfill functions and for the design and operation of leachate treatment facilities, facilitates risk analysis of leachate impact on the environment should liners leak, permits comparison of the impact of alternative landfill design or operating protocol on the environment, and discloses the interaction of leachate parameters. Critical parameters for monitoring and control are described in more detail in Chapter 8.

Material is removed from the waste mass via mechanisms that include leaching of inherently soluble material, leaching of soluble products of biological and chemical transformation, and washout of fines and colloids. The characteristics of the leachate are highly variable depending on the composition of the waste, rate of water infiltration, refuse moisture content, and landfill design, operation, and age. These variations are demonstrated in Table 2.1, where ranges in concentrations of significant leachate components are presented as a function of stabilization phase.

Organic contaminants of leachate are primarily soluble refuse components or decomposition products of biodegradable fractions of waste. Organic compounds detected in nineteen MSW landfill leachates or contaminated groundwater plumes emanating from landfills included organic acids, ketones, aromatic compounds, chlorinated aromatic compounds, ethers, phthalates, halogenated aliphatic compounds, alcohols, amino-aromatic compounds, nitro-aromatic compounds, phenols, heterocyclic compounds, pesticides, sulfur substituted aromatic compounds, polyaromatic hydrocarbons, polychlorinated biphenyls, and organophosphates.[17]

The class of organic compounds found at highest concentration in leachates is generally VOAs produced during the decomposition of lipids, proteins, and carbohydrates.[18-19] Aromatic hydrocarbons, including benzene, various xylenes, and toluene, are also frequently found at lower concentrations.[19-20] These compounds were considered to be constituents of gasoline and fuel oils. Sawney and Kozloski[21] reported that the presence

TABLE 2.1

Landfill Leachate Concentration Ranges as a Function of the Degree of Landfill Stabilization[16]

Parameter	Phase II Transition	Phase III Acid Formation	Phase IV Methane Formation	Phase V Final Maturation
BOD, mg/l	100–10,000	1000–57,000	600–3400	4–120
COD, mg/l	480–18,000	1500–71,000	580–9760	31–900
TVA, mg/l as Acetic Acid	100–3000	3000–18,800	250–4000	0
BOD/COD	0.23–0.87	0.4–0.8	0.17–0.64	0.02–0.13
Ammonia, mg/l–N	120–125	2–1030	6–430	6–430
pH	6.7	4.7–7.7	6.3–8.8	7.1–8.8
Conductivity, (μmhos/cm	2450–3310	1600–17,100	2900–7700	1400–4500

of the more soluble, less volatile aromatic components of gasoline in the leachate suggested that the more volatile components were being gas stripped from the landfill. Nonvolatile classes of compounds such as phenolic compounds may be degradative byproducts of lignin. A small complex fraction found in several leachates contained nicotine, caffeine, and phthalate plasticizers.[18] Oman and Hynning[22] observed that a total of 150 different organic compounds have been identified in multiple studies, however only 29 were identified in more than one, concluding that leachate composition was quite site specific.

The dominant organic class in leachate shifts as the age of the landfill increases due to the ongoing microbial and physical/chemical processes within the landfill. An investigation of leachates obtained from landfills operated from one to 20 years found that the relative abundance of high molecular weight humic-like substances decreases with age, while intermediate-sized fulvic materials showed significantly smaller decreases.[23] The relative abundance of organic compounds present in these leachates was observed to decrease with time in the following order: free VOAs, low molecular weight aldehydes and amino acids, carbohydrates, peptides, humic acids, phenolic compounds, and fulvic acids.

A variety of heavy metals are frequently found in landfill leachates including zinc, copper, cadmium, lead, nickel, chromium, and mercury.[24] Again, these metals are either soluble components of the refuse or are products of physical processes such as corrosion and complexation. In several instances heavy metal concentrations in leachate exceed U.S. Toxicity Characteristic Leaching Procedures standards.

Heavy metal concentrations in leachate do not appear to follow patterns of organic indicators such as COD, BOD, nutrients, or major ions.[24] Heavy metal release is a function of characteristics of the leachate such as pH, flow rate, and the concentration of complexing agents.

Leachate Treatment and Disposal

Leachate that is collected and removed from a landfill must be managed in a suitable fashion. This entails some type of treatment process, either at the landfill site or at a treatment facility offsite. Treated leachate, as with any treated wastewater, must meet appropriate regulatory limits for discharge to the environment.

Perhaps the simplest approach to managing leachate involves discharge to a local wastewater treatment plant. If a sewer connection is located at the landfill site, leachate may be directly discharged from the leachate storage facility. Most landfills, however, are located in less populated areas, where direct sewer connections are often not available. In this case, leachate may be hauled to a local wastewater treatment facility by means of tanker trucks. Landfill operators often encounter resistance by wastewater treatment plants because of the possible impact of leachate on the treatment process. Although the volume of leachate discharged normally is much less than the volume of wastewater, leachate exhibits large variations in quantity and quality, and at times contains high concentrations of potentially disruptive chemicals. Pretreatment and flow equalization are often required.

If the cost of transporting and treating leachate offsite is prohibitive, a treatment facility may be constructed onsite. Much work has been conducted regarding the various treatment alternatives for landfill leachate.[16] Processes that have been evaluated include biological treatment, both aerobic and anaerobic, and physical-chemical processes such as activated carbon, chemical oxidation and precipitation, and reverse osmosis. A key point to consider in the design of a leachate treatment facility is that leachate quality and volume may vary greatly over the life of the landfill. Flow equalization is therefore important. Also, the types of constituents found in leachate differ from typical domestic wastewater. In addition to organic compounds that require biological treatment, leachate contains inorganic dissolved solids (chloride, sodium) which experience limited removal by biological treatment. Many full-scale leachate treatment operations therefore include both biological and physical-chemical unit operations. Natural treatment operations such as wetlands have also been used in some cases to polish leachate.

Characteristics and Generation of Landfill Gas

As previously stated, when solid waste decomposes, significant portions of organic wastes are ultimately converted to gaseous end-products. The rate of gas production is a function of refuse composition, climate, moisture content, particle size and compaction, nutrient availability, and buffering capacity. Reported production rates vary from 0.12 to 0.41 m^3/kg (3,800 to 13,000 ft^3/ton) dry waste.[16] Production rates and gas composition also follow typical stabilization phases (Fig. 2.6), with peak flow rates and methane content occurring during the methanogenic phase. Landfill gas is typically 40 to 60 percent methane, with carbon dioxide and trace gases such as hydrogen sulfide, water vapor, hydrogen, and various volatile organic compounds comprising the balance.

Because of their high vapor pressures and low solubilities, many toxic volatile organic compounds (VOCs) are observed in landfill gas. In a report by the State of California Air Resources Board,[25] the average surface emission rate of hazardous chemicals from landfills was estimated to be 35 kg/million kg of refuse. Landfill gas trace constituents include halogenated aliphatics, aromatics, heterocyclic compounds, ketones, aliphatics, terpenes, and alcohols and have been characterized by several researchers.[26-29]

Landfill Gas Control

The production of LFG at a MSW landfill presents a new set of design challenges in regard to the control of gas migration, the collection of gas flow from the fill, and the treatment and potential reuse of the gas. As mentioned previously, gas collection is conducted to minimize emissions to the atmospheric for health and safety concerns, aesthetics, and to minimize atmospheric degradation. While states such as California have required gas control for a number of years, federal regulations were only recently promulgated to require control of gas emissions from large landfill sites. Typical gas collection systems utilize vertical wells placed within the landfill at the time of closure. These wells are similar to those used for groundwater and consist of perforated pipe surrounded by a permeable media such as gravel. A typical vertical gas recovery well is indicated in Fig. 2.7.

While gas generated within the landfill will migrate toward a well due to the pressure difference between the landfill interior and the atmosphere, passive venting does not always result in large collection efficiencies. An alternate approach is to place a vacuum on the well, thus creating a greater potential to remove gas from the landfill. This normally is accomplished by connecting individual wells to a pipe network that is in turn connected to a mechanical blower. The blower induces a vacuum in the manifold and the wells, extracting gas from the landfill interior. The goal of an active landfill gas collection system is to remove the maximum amount of gas possible from the waste, thus minimizing migration to the atmosphere. The applied vacuum must not be so great, however, that air is drawn into the landfill. The presence of oxygen may result in a landfill fire, the methanogenic microbial activity may be suppressed, and the treatment or reuse operation may suffer.

The most common treatment process for LFG is to combust the gas using a controlled flare, resulting in destruction of the methane and the trace organic contaminants. The conversion of the methane to a recoverable product is another possibility. LFG recovery operations that have

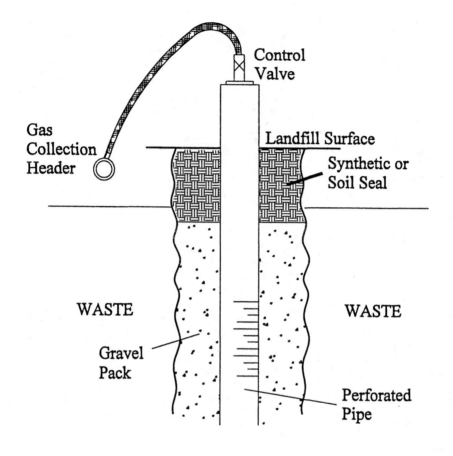

Figure 2.7 Typical landfill gas collection well.

been performed include cleanup and sale to existing natural gas pipelines, the operation of gas-fired electrical generators, and the use of the gas as a vehicle fuel.

Waste Decomposition and Landfill Settlement

The material that remains in a landfill after the waste has been stabilized consists of any non-biodegradable waste (metal, plastic, glass) as well as residual biodegradable materials. During the process of landfill stabilization, waste mass is lost through the production of LFG. The resulting landfill mass will therefore settle, decreasing the volume of landfilled material. As a result of the heterogeneous nature of MSW and the different degrees of stabilization that may occur in a single landfill unit, waste settlement is rarely uniform. The differential settlement which results must be considered in the design of landfill gas collection manifolds and the surface capping system. The landfill must be carefully monitored after closure to maintain the integrity of the capping system.

■ LANDFILL OPERATION STRATEGIES

Current regulations for MSW landfills emphasize the landfill operation strategy of containing and removing leachate before it enters the environment, and limiting the original generation of the leachate. The presence and movement of moisture in the landfilled waste is one of the most important factors controlling waste degradation and landfill stabilization. By minimizing the amount of water that enters the landfill, the stabilization of the waste is minimized. This strategy has been termed the "dry tomb" approach. An alternate strategy, which is the subject of this book, is to actively encourage the stabilization of the landfill through technologies such as leachate recirculation.

3 LANDFILL BIOREACTOR STUDIES

■ LABORATORY SCALE STUDIES

Many laboratory scale studies have been conducted to investigate the effects of leachate recirculation on leachate quality, waste stabilization, waste settlement, gas production, attenuation of heavy metals, and other factors. Several of these studies are described in this chapter and are introduced chronologically. The laboratory-scale studies described below are by no means all of the pertinent research reported in the literature. Other studies include those of Mata-Alvarez and Martina-Verdure,[30] Barlaz, et al.,[31] Buivid, et al.,[32] Chian and DeWalle,[33] and Klink and Ham.[34]

Moisture content, pH, temperature, availability of macro- and micro-nutrients and the presence of suitable microorganisms are the main parameters controlling the process of landfill stabilization and, therefore, are the parameters typically manipulated in the laboratory studies. Moisture content can be controlled by the addition of regulated quantities of water and/or leachate. The pH can be controlled by adding buffering compounds. Macro- and micro-nutrients are usually present in sufficient quantities in the waste and do not act as limiting factors in the stabilization process, hence nutrients usually are not added. Presence of suitable microorganisms responsible for stabilization can be ensured by adding anaerobically digested sludge in which acclimated anaerobic and facultative microorganisms are present. The impacts of these parameters have been investigated numerous times and will be discussed as appropriate in the review of each study.

Georgia Institute of Technology Experiment I[2]
One of the first experiments on leachate recirculation was conducted at the Georgia Institute of Technology in the mid-1970s and supported by the U.S. Environmental Protection Agency (EPA). This experiment, which conclusively proved the effectiveness of leachate recirculation on waste stabilization, is described below.

Experimental Setup
Four test columns, each 0.9 m (3 ft) in diameter, were filled with 3 m (10 ft) of MSW and compacted to a density of 357 kg/m³ (535 lb/yd³). A 0.76-m (2.5-ft) soil cover was placed over the compacted coarsely

ground waste mass. The first column was a control cell with no leachate recirculation (single-pass cell). The second column was subjected to leachate recirculation only. In the third column, leachate recirculation was coupled with pH control (using NaOH) to maintain near neutral conditions. Leachate recirculation with pH control in addition to initial seeding (with wastewater sludge) was used for the fourth column. Leachate recirculation was accomplished by pumping through a distribution system located on top of the waste but below the cover soil. Approximately 945 l (250 gal) of water were added to all of the cells to produce leachate immediately. Leachate was drained to separate sumps from which samples were collected and analyzed at regular intervals for 1100 days after the start of the experiment.

Results
For the control column, Chemical Oxygen Demand (COD) increased rapidly and peaked at about 19,000 mg/l after about 200 days and declined gradually to about 4,000 mg/l after 1,000 days. Total volatile acids (TVA) concentration showed a similar pattern, peaking at about 10,000 mg/l after 200 days and declining to 2,000 mg/l after 1,000 days. The pH did not vary significantly and was in the acidic range of 5.0 to 6.5 throughout 1100 days of monitoring.

For the second column, with leachate recirculation only, the variation of COD, TVA, and pH with time was significantly different from the control cell. COD increased more rapidly than the control column, peaking at 11,000 mg/l after 100 days. After 150 days, COD decreased to a low of 4,000 mg/l and then decreased gradually to about 250 mg/l by day 500. TVA concentration variation followed the pattern of COD variation, peaking to 6,000 mg/l at 200 days, but increased again at 250 days, reaching 1,500 mg/l at 400 days and declined gradually to less than 200 mg/l by day 700. The pH remained around 5.0 for the first 200 days. As the COD and TVA concentrations dropped, pH increased to about 7.0 after 500 days and remained in the neutral region thereafter.

Column 3 with leachate recirculation and pH control, showed a more rapid decrease in TVA and COD concentrations than the second column. Both COD and TVA peaked after 150 days at 10,000 mg/l and 5,000 mg/l, respectively. After 200 days, COD concentration had decreased to less than 500 mg/l and TVA concentration to less than 250 mg/l. Due to the addition of buffers, pH was near neutral throughout the monitoring period.

The fourth column, with leachate recirculation, pH control and seeding, did not perform better than the pH-controlled leachate recirculating column (Column 3). Peak COD and TVA concentrations were 9,000 mg/l and 5,000 mg/l respectively, both occurring at about 60 days. While the TVA concentration declined to near zero at 400 days, COD concentration reached near zero at 650 days. Initially, pH dropped to nearly 5, but due to buffering, was in the neutral region after day 150 and throughout the remaining monitoring period.

Conclusions

The following important observations can be made from the results of this laboratory simulation:

- Leachate recirculating columns produced low COD/TVA leachates in a shorter time period as opposed to a more gradual decline in the control cell.

- The peak COD and TVA concentrations in the leachate recirculated columns were less than the control column.

- pH remained more neutral in the leachate recirculated column than the control column.

- pH control and leachate recirculation gave the best performance with rapid decline in COD and TVA concentrations.

- Inoculation with wastewater sludge did not accelerate the degradation process.

The most significant inference that can be drawn from this experiment is that leachate recirculation accelerates the waste degradation process as characterized by a rapid decrease in the COD and TVA concentrations. Although this was one of the first experiments in the area, it convincingly proved the effectiveness of leachate recirculation.

University of Louisville Experiment[35]

A laboratory-scale study, supported in part by the Illinois Institute on Environmental Quality, was conducted to demonstrate the advantages of leachate recirculation and to investigate the feasibility of leachate recirculation systems in providing leachate treatment. An additional objective was to determine of the effect of leachate pH and nutrient control on biological stabilization of shredded and unshredded refuse.

Experimental Setup

Four 0.9-m (3-ft) diameter steel test cells, equipped with leachate collection and redistribution (except the control cell) systems, were set up and filled to 2.4 m (8 ft) with domestic refuse compacted to a density of 240 kg/m^3 (400 lbs/yd^3). Test Cell 1 was the control cell with unshredded refuse. Cell 2, with leachate recirculation, contained shredded refuse with 70 percent moisture content and pH control (using NaOH). Cell 3 was operated similarly to Cell 2 except that it contained unshredded refuse. Test Cell 4 had nutrient control (phosphorus and nitrogen addition) and was similar to Cell 3 in all other aspects. Approximately 11 m^3 (200 gal) of water were added to start leachate production in Cells 2, 3, and 4. Tap water, equivalent to the total daily rainfall, was routinely added to all of the cells. Leachate from all the cells was analyzed periodically for 514 days.

Results

The data obtained showed that leachate recirculation resulted in marked decrease in the concentrations of TVA, Biochemical Oxygen Demand (BOD), COD, and Total Organic Carbon (TOC) in Cells 2, 3, and 4 in comparison to Cell 1. Cell 1 appeared to be acid stuck, maintaining high levels of TVA throughout the test period, although pH remained at neutral levels. Because all three leachate recirculating cells produced consistent leachate quality, nutrient control was found not to have any significant impact on leachate quality nor did shredding of waste increase the rate of stabilization. The BOD/COD ratios were observed to be close to those reported for municipal wastewater (between 0.6 to 0.8) in all four cells, indicating the biological treatability of leachate in wastewater treatment plants should offsite treatment of leachate be required.

Conclusions

Based on the analytical data obtained for all the cells, the following conclusions were drawn:

- Leachate recirculation with pH control established anaerobic biological population in the fill rapidly.

- Nutrient control did not have any significant effect on stabilization of organic content of the refuse.

- Shredding did not have any effect on biological stabilization of the refuse.

- Leachate recirculation with pH control lead to accelerated biological stabilization of the organic content of the refuse thereby reducing the time required for ultimate site use (land reclamation).

- Leachate recirculation with pH control lead to significant reductions in BOD, COD, and TOC.

- Leachate recirculation with pH control can be used as an effective leachate treatment process.

Federal Republic of Germany Experiment[36]

Experimental Setup

Test cell experiments on leachate recirculation were conducted in the former Federal Republic of Germany by Doedens and Cord-Landwehr.[36] Four steel cylindrical test cells, each 1.5 m (4.9 ft) in diameter were installed and filled with 1.35 m (4.4 ft) of shredded waste compacted to a density of to 20,800 to 23,200 kg/m^3 (1300 to 1450 lb /yd^3) wet weight to simulate the actual density achieved in field by compactors. The waste volume in each cell was 2.4 m^3 (85 ft^3) and the initial moisture content was 24 to 31 percent. All of the cells were air tight, temperature controlled (35°C), and equipped with leachate redistribution system and meters to measure leachate and gas flows.

Test Cell 1 received rainwater equivalent to 660 mm/yr (26 in/yr) of precipitation and all the leachate generated was redistributed to the cell. The remaining Cells (2, 3, and 4) received rainwater equivalent to 330 mm/yr (13 in/yr) precipitation (half that of Cell 1). Test Cell 2 was the control cell receiving rainwater only with no leachate recirculation (single-pass cell). Test Cell 3 received rainwater (330 mm/yr) (13 in/yr) and all the leachate generated was fed back to the cell along with rainwater. The fourth cell was initially brought to field capacity with leachate from a stabilized landfill. Thereafter, the cell received 330 mm/yr (13 in/yr) rainwater in addition to all of the leachate generated.

Results
Test Cell 1 showed a more rapid decline in the COD concentration than any other cell. Test Cell 2, the control cell, took longer than any other cell to reduce COD concentration to levels comparable to other cells. After about 300 days from the start of the experiment, Test Cell 1 had the lowest COD concentration and Test Cell 2 had the highest concentration. Test Cells 3 and 4 produced approximately the same COD concentrations, which were higher than Test Cell 1 and lower than Test Cell 2. However, since varying quantities of water were added to each of the cells, the total mass loads emitted from the cells should be considered rather than concentrations. For example, although the COD concentration is less in Test Cell 1, the total load emitted would be more because the cell received twice the amount of water than that of other cells. Considering the loads withdrawn from the test cells during the test period, excluding the loads in recycled leachate, Test Cell 3 gave the best result with the lowest mass emission during the test period of COD, BOD, chloride, zinc, lead, and cadmium compared to remaining cells.

Cell 4 had the highest emission loads of COD, BOD, chloride, zinc, lead, and cadmium due to its initial saturation with stabilized landfill leachate. Cell 1 (with leachate recirculation) performed better than Cell 2 (control cell) with respect to emission loads of COD, BOD, chloride, zinc, lead and cadmium. Control Cell 2 (single-pass cell) produced more gas than any other cell (114 m³/t dry matter (3650 ft³/ton)), followed by Cell 3 at 111 m³/t dry matter (3560 ft³/ton) and Cell 4 at 105 m³/t dry matter (3360 ft³/ton). Cell 1 produced significantly less gas than any other cell (50 m³/t dry matter [1600 ft³/ton]). Excluding Cell 4, which was initially saturated with leachate, the two leachate recirculating cells performed better (with respect to leachate quality) than the control (single-pass) cell. The most significant inference from this experiment is that leachate recirculation lowers the emitted loads of COD, BOD, chloride, and metals.

Newcastle University Experiment[37]

Experimental Setup

Lysimeter studies were conducted at Newcastle University, U.K. to investigate the effect of leachate recirculation on different types of wastes and under various operating conditions. Four lysimeters, each 0.5 m in diameter, were filled with various types of domestic wastes and compacted to different densities. Depth of the columns was not indicated. The lysimeters were equipped with leachate distribution piping, leachate monitoring systems, and gas monitoring systems. Lysimeter 1 was filled with fresh crude domestic waste with a moisture content of 61 percent, and a density of 383 kg/m³ (648 lbs/yd³) while lysimeter 2 was filled with the same type of waste and moisture content but was compacted to a density of 418 kg/m³ (702 lbs/yd³). lysimeter 3 consisted of shredded domestic refuse with a moisture content of 44 percent and density of 306 kg/m³ (513 lbs/yd³). Lysimeter 4 consisted of aged domestic refuse with a moisture content of 85 percent and density of 550 kg/m³ (34 lbs/ft³). Lysimeter 2 was operated under saturation conditions while all others were operated with free draining leachate. The lysimeters were monitored over a period of 400 days.

Results

Leachate from lysimeter 3 with shredded refuse and the lowest density, showed a marked decrease in COD concentration within a shorter time than lysimeters 1 and 2. Lysimeter 4, with aged refuse, had a very low COD leachate throughout the test period. Lysimeters 1 and 2 leachates had approximately the same COD concentration that was significantly higher than lysimeter 3. Similar results were observed with respect to TOC concentrations. Results were similar with respect to TVA and BOD concentration variations, with lysimeter 3 showing a dramatic decrease within a shorter time when compared to lysimeters 1 and 2 and lysimeter 4 having low concentrations throughout. Clearly, lysimeter 3 performed better than any of the other lysimeters except for lysimeter 4 containing aged refuse.

Conclusions

The following conclusions can be drawn from the results of this experiment:

- Shredding of refuse increases the degradation rate thereby producing better quality leachate in a shorter period of time.

- Lower density helps in increasing the degradation rate.

- Operation under saturation conditions does not result in any benefits with respect to waste degradation, but rather may lead to higher strength leachates.

Georgia Institute of Technology Experiment II[38]

Another EPA sponsored laboratory-scale experiment was conducted at the Georgia Institute of Technology to study the influence of leachate recirculation and single-pass operation on selected inorganic and organic priority pollutants codisposed with shredded municipal solid waste.

Experimental Setup

The experimental setup consisted of 10 simulated landfill columns (steel cylinders) 0.9 m (3 ft) in diameter and 3 m (10 ft) in height. Each column was lined with 30-mil (0.076-cm) High Density Polyethylene (HDPE) liner and equipped with leachate collection and moisture/leachate distribution system. Each column was filled with 378 kg (832 lb) of shredded municipal solid waste. Organic and/or inorganic priority pollutants were added to the columns (except control columns) in varying doses. Metal sludges (constituting inorganic priority pollutants), prepared by mixing sawdust and metal oxides to obtain low, medium or high loadings were added to the columns. Two identically loaded columns, one with single-pass operation and another with leachate recirculation, constituted a pair of columns giving a total of five pairs of columns as shown in Table 3.1.

Results

Leachate and gas samples were routinely analyzed for three years for admixed pollutants and other parameters. Overall, significantly greater volumes of gas at higher rates were produced in recycling columns vs. single-pass columns (averaging 42 m³ (1500 ft³) and 8.1 m³ (290 ft³) respectively) indicating an enhanced degree of waste stabilization. A

TABLE 3.1

Column Loading Characteristics[38]

Column	Operation	Initial Loading Height (cm)	Compact Density (kg/m³)	Inorganic Pollutant Added[a]	Organic Pollutant Added[b]
1 CR	Recycle	29	313	None	None
2 C	Single pass	30	301	None	None
3 O	Single pass	29	309	None	Yes
4 OL	Single pass	28	327	Low	Yes
5 OM	Single pass	30	305	Medium	Yes
6 OR	Recycle	28	317	None	Yes
7 OLR	Recycle	29	309	Low	Yes
8 OH	Single pass	30	305	High	Yes
9 OMR	Recycle	29	313	Medium	Yes
10 OHR	Recycle	31	293	High	Yes

[a] Low: Cd = 35 g; Cr = 45 g; Hg = 20 g; Ni = 75 g; Pb = 105 g; Zn = 135 g, Medium: Low doubled, High: Medium doubled.

[b] 120 g each of 12 different organic compounds.

prolonged acid phase was experienced in all columns requiring the addition of anaerobically digested sludge to initiate methane fermentation. During this time TVA concentrations reached 15,000 mg/l in recycle columns and 5 to 10,000 mg/l in single-pass columns. Leachate strength (as indicated by COD and TVA concentrations) decreased and gas production increased only after the onset of the methanogenesis phase after approximately 720 days. Thereafter, there was a dramatic decrease in COD and TVA concentrations along with an increase in pH and gas production in leachate recirculated columns with more moderate changes in single-pass columns. Gas production, also stable initially, dramatically increased with the onset of methane fermentation.

Conclusions
Priority pollutants caused a delay in the stabilization process in recycling columns and total inhibition in single-pass columns as demonstrated by lower pH values, gas production, and higher TVA concentrations relative to the control column. With respect to heavy metals, both recirculating and single-pass columns were capable of assimilating added pollutants, however recycle columns demonstrated greater capacity. Single-pass and leachate recycle columns exhibited little difference with respect to the release patterns of organic priority pollutants like dibromomethane, trichloroethane, and nitrobenzene. Leachate containment by recirculation, in addition to providing *in situ* leachate treatment, resulted in efficient conversion to gas of many organic leachate constituents that otherwise would have been washed out.

■ PILOT-SCALE BIOREACTOR STUDIES

A number of pilot-scale studies were conducted in the U.S. and other European countries to investigate the effects of leachate recirculation on landfill stabilization, leachate quality, landfill gas production and other parameters. Moisture content, waste density, type of waste, pH, temperature, nutrients, and seeding (addition of sludge) typically were controlled and manipulated in the pilot- scale studies. Several pilot-scale studies are briefly described below. A detailed compilation of test cell data has been provided elsewhere.[39]

Sonoma County, California[40,41]

Experimental Setup
A pilot-scale landfill project was started in 1972 at Sonoma County, California, to study the effect of moisture on the rate of refuse stabilization, and on leachate quantity and quality. Five large-scale field test cells, each 15 m (49 ft) by 15 m (49 ft) by 3 m (10 ft), were constructed, filled with municipal solid waste compacted to 630 kg/m³ (1000 lb/yd³), and subjected to different operating conditions (moisture regimes). Each cell

served a specific purpose and was operated accordingly as shown in Table 3.2. The cells were monitored for 900 days to evaluate leachate quality, gas composition, and settlement.

Cell A was the control cell, Cell B was initially brought to field capacity, Cell C received water at a rate of 3.8 m³/d (1000 gal/d), Cell D was subjected to leachate recirculation, and Cell E was initially inoculated with septic tank pumpings. After construction, Cells A, B, and E received moisture only from infiltrating rainwater. The rate of leachate recirculation (in Cell D) varied from 1.9 to 19 m³/day (500 to 5,000 gal/day). The leachate organic strength for Cells C and D with water input gradually declined. Leachate recirculation provided a more rapid decline in COD concentration than all other cells. The data from Cell D on gas composition also indicated an increased rate of biological stabilization.

Conclusions
Based on the performance evaluation of Cell D and comparison with other cells, the following specific conclusions with respect to leachate recirculation were made:

- Leachate recirculation significantly increased the rate of establishment of anaerobic microbial population within the fill as suggested by gas quality.

- Leachate recirculation increased the rate of biological stabilization of the organic fraction of the refuse (as evidenced by reductions in BOD, COD, and TVA concentrations in leachate).

- Settlement was enhanced by liquid flow and accelerated microbial activity with leachate recirculation (20 percent reduction in height for leachate recirculating cell vs. 7.6 percent for remaining cells).

- Continual flow-through of water increased rate of stabilization but created large volumes of leachate requiring *ex situ* treatment.

- The addition of septic tank pumpings (inoculum) accelerated acid fermentation and was not beneficial in the absence of pH control and leachate recirculation.

- In the presence of leachate recirculation, the landfill acted as an anaerobic digester in treating leachate and therefore was found to be the most feasible and beneficial management strategy utilized in the study.

Georgia Institute of Technology Study[42]
Pilot-scale investigations were conducted at the Georgia Institute of Technology to study the effects of leachate recirculation and to augment the results of initial laboratory-scale studies also conducted at Georgia Institute of Technology.

TABLE 3.2

Test Cell Moisture Conditions, Sonoma County[40]

Cell	Purpose	Initial Liquid Condition	Liquid Used in Initial Conditioning	Daily Liquid Application, m³/d[a]	Liquid Used for Daily Liquid Application
A	Control cell	None	None	None	None
B	High initial water content	Field capacity	Water	None	None
C	Continual addition of water	None	None	0.76–3.8	Water
D	Leachate recirculation	None	None	1.9–3.8	Leachate
E	Microbial seeding	Field capacity	Septic tank pumpings	None	None

[a] Other than infiltrating liquid

Experimental Setup
Two simulated concrete landfill cells, each 3 m (10 ft) by 3 m (10 ft) by 4.3 m (14 ft) deep were constructed and filled with 3 m (10 ft) of shredded municipal solid waste compacted to a density of 319 kg/m^3 (537 lb/yd^3). One cell was left open to incident rain and the other was sealed completely to permit gas collection and eliminate evaporation. The cells were allowed to reach field capacity over a year's time. The sealed cell received tap water equivalent to rainfall received by the open cell through the distribution system. Both of the cells were equipped with leachate collection and distribution systems.

Results and Conclusions
Nominal leachate production began on September 1, 1977, about a year after the completion of filling operations (August 13, 1976). Available leachate was recycled on a weekly basis. Daily recirculation of 760 l/day (200 gal) per cell was started on day 208. On day 346, the quantity of leachate recirculated was reduced to percolation capacity of the open cell. Recycling on alternate days was started on day 401 and was gradually reduced to 150 to 190 l/day per cell (40 to 50 gal/day).

Leachate samples were collected from both cells at regular intervals and analyzed for various parameters including BOD, COD, TOC, TVA, pH, phosphorus, chloride, and selected metals. Gas samples were also analyzed for carbon dioxide, nitrogen, methane, and hydrogen. TOC, COD and BOD, the critical parameters indicative of pollution potential of leachates, displayed similar pattern of decrease in concentrations for both cells. Data obtained during initial periods of weekly recirculation were somewhat erratic, more so for the sealed cell due to uneven distribution of TOC, COD and BOD after initiation of daily circulation. Although there was not much difference in concentrations of TOC, COD and BOD toward the end of the test period (520 days after leachate production began) concentrations were lower in the sealed cell than the open cell. TVA concentrations in both the cells initially increased and later decreased to levels below detection. The sealed cell provided a more congenial environment to methane formers by excluding oxygen. Gas production in the sealed cell increased to a high of 0.64 m^3/d (23 ft^3/d) with a methane content of 57 percent, coinciding with the decrease in TVA concentration and increase in pH. Gas production decreased to 0.01 to 0.02 m^3/d (0.35 to 0.7 ft^3/d) by the end of the test period indicating that most of the readily available organics in the leachate were converted to gas within three months. Based on the differences in chloride concentrations between the two cells, moisture loss in the open cell due to evaporation was estimated between 20 to 30 percent of the incident rainfall.

This pilot-scale study supported the results of the previous laboratory-scale studies and demonstrated the advantages of leachate recirculation in rapid stabilization of readily available organic constituents accompanied with increased gas production rate in short time period.

Mountain View Landfill, California[43,44]

The objective of the pilot scale demonstration project conducted at Mountain View Landfill, California, was to study the effectiveness of the methods used to enhance methane gas generation. The key factors controlled and manipulated were moisture addition, buffer, inoculation, and leachate recirculation. However, as testified by refuse analysis, there was no effective control over moisture content due to excessive water infiltration into the test cells.

Experimental Setup

Six separate cells (designated A through F) were constructed for the study purpose. Each cell was approximately 31 m by 31 m (100 ft by 100 ft) and contained 14 m (47 ft) of waste. Cell F was the control cell and partial leachate recirculation was provided for Cell A. The other cells were not subjected to leachate recirculation. Composition of the cells at the beginning of the project (June 1981) is shown in Table 3.3. The cell characteristics and monitoring results after 1597 days (through December 1985) are presented in Table 3.4.

Results

The results with respect to leachate recirculation were not in agreement with other studies. The total gas production rates were lower than the rates obtained in other lysimeter studies. Cell D, with no sludge addition, yielded the maximum amount of gas even though the moisture content (as compared to other cells) was low in the beginning and lowest at the end of the study period. Cell A (with partial leachate recirculation) had the highest moisture content at the end of the study period, yet produced less gas then Cell C (Cell C was identical to Cell A except without leachate recirculation). However, leachate recirculation resulted in faster stabilization as indicated by volatile solids content, cellulose content, carbon-to-nitrogen ratios, and carbon-to-phosphorus ratios presented in Table 3.5.

Conclusions

There were numerous discrepancies between measured and calculated gas production rates (based on loss of volatile acids). Except for Cell D, which had negligible infiltration, the measured gas production rates for all the other cells were less than calculated values. Cells A and B with highest water infiltration had lowest measured gas production rates. Calculated gas production values indicated that high moisture content (and possibly leachate recirculation) and the addition of sludge increased methane production. Inconsistencies in gas production data were attributed to gas leakage from test cells.

On the basis of refuse analysis from cells A, B, D and F, the following conclusions were made:

Table 3.3

Cell Composition After Construction, Mountain View[44]

Cell Component	A	B	C	D	E	F
Additions[a]	sbrw	sb	sbw	b(w)	s(w)	none
Dry refuse solids (million kg)	4.88	5.42	4.81	6.00	4.96	5.64
Refuse associated water (million kg)	1.63	1.81	1.60	2.00	1.65	1.88
Porosity (%)	50	49	50	49	51	48
Dry sludge solids (million kg)	0.16	0.13	0.07	0	0.06	0
Sludge associated water (million kg)	0.88	0.73	0.38	0	0.31	0
Sludge in place (million kg)	1.03	0.86	0.44	0	0.37	0
Buffer (million kg)	0.010	0.010	0.009	0.010	0	0
Precipitation (million kg)	0.13	0.14	0.13	0.14	0.14	0.14
Total dry solids (million kg)[b]	5.05	5.56	4.89	6.01	5.02	5.64
Total water (million kg)	4.34	1.95	3.81	2.38	2.34	2.02

[a] Additions: s = anaerobic digester sludge; b = buffer (calcium carbonate), approximately 9070 kg; w = water, 1700 m^3; (w) = water, 235–238 m^3; r = recirculation of leachate.

[b] Excluding buffer which equaled approximately 9070 kg.

TABLE 3.4

Cell Construction Characteristics and Monitoring Results (1597 Days), Mountain View[44]

Cell Component	Cell					
	A	B	C	D	E	F
Additions[a]	sbrw	sb	sbw	b(w)	s(w)	none
Moisture content at construction (%)[b]	46	32	44	28	32	26
Moisture content at conclusion (%)	69	54	50	33	45	40
Specific landfill gas yield (m³/dry kg)	0.08	0.07	0.09	0.16	0.07	0.14
Specific methane yield (m³/dry kg)	0.04	0.04	0.05	0.09	0.04	0.08
Conversion (% of ultimate)[c]	19	17	22	40	16	33
Average gas production rate (m³/dry kg)	0.02	0.02	0.02	0.04	0.02	0.03
Total landfill gas produced (thousand m³)	314	275	334	748	238	631
Average cell settlement (m)	2.0	2.2	2.3	1.3	2.3	1.9

[a] Additions: s = anaerobic digester sludge; b = buffer (calcium carbonatge), approximately 9070 kg; w = water, 1700 m³; (w) = water, 235–238 m³; r = recirculation of leachate.
[b] After water addition.
[c] Calculated ultimate yield = 0.23 m³ methane/dry kg refuse.

TABLE 3.5

Refuse Chemical Analysis Summary, Mountain View[44]

| | Cell[a] | | | |
	A	B	D	F
Sampling interval (m)	0–7.6	0–8.5	0–9.8	0–11.9
Moisture content (%, wet weight basis)	68.6	54.1	33.3	40.0
Volatile solids content (%)	31.8	43.1	50.7	43.5
Cellulose (%)	16.3	25.6	32.8	26.6
Lignin (%)	13.4	14.0	13.6	14.2
Carbon to nitrogen ratio	13:1	20:1	26:1	27:1
Carbon to phosphorus ratio	6593:1	945:1	1345:1	1169:1

[a] Samples were not obtained from Cells C and E.
Note: All parameters except moisture content measured on dry basis.

- Cells with higher moisture content, sludge addition, less settlement and lower internal temperatures had lower measured gas production rates.

- The calculated average yearly methane gas production rates based on loss of volatile solids indicated that a higher moisture content, addition of sludge and leachate recirculation enhanced methane gas generation which is in contradiction to measured data. The contradiction was attributed to gas leaks and water infiltration.

- The relationship between moisture infiltration and measured gas production rates (i.e., cells with higher infiltration had lower measured gas production rates) suggests that the pathways of moisture infiltration and gas escape might have been the same.

Binghamton, New York[45]

This study conducted for the New York State Energy Research and Development Authority (NYSERDA) was one of the first pilot-scale experiments to investigate the enhancement of landfill gas production by leachate recirculation. The objective of the study was to examine landfill gas production while varying the key parameters controlling anaerobic digestion, namely, moisture content, pH, temperature, and nutrients. Nutrients were controlled by varying the quantity of wastewater treatment plant sludge added to the waste. The pH was controlled by the addition of buffers.

Experimental Setup
Nine pilot-scale Polyvinyl Chloride (PVC)-lined landfill cells (designated Cell No. 1 through Cell No. 9), 6.4-m (21-ft) deep with 17 m (57 ft) by 23 m (75 ft) foot print, were set up at Nanticoke Landfill, Binghamton, NY. Each cell was equipped with leachate collection, leachate/moisture

distribution, and gas collection and metering systems. The first cell was a control cell, the second cell received moisture only (no addition of sludge or buffer), and the third cell received moisture and buffer (lime). The fourth cell received anaerobically digested sludge but no buffer or water. The remaining three cells received both sludge and buffer in varying amounts. Each cell was an encapsulated system separated from other cells. Although nine cells were constructed, only seven were operated. (Cells 7 and 8 were not operated). The cells were monitored for a period of two years.

Results and Conclusions
Based on gas monitoring data, it was clear that the cells with sewage sludge yielded significantly higher quantities of gas than the cells without sewage sludge. Also, the methane content was higher in the cells with sludge than the cells without sludge. Table 3.6 presents data on cell composition; Table 3.7 provides leachate quality data and gas production.

Leachate quality in high gas yielding cells (Cells 4, 5, and 6) was better (lower in strength with respect to COD, TVA and alkalinity) than the low gas yielding cells (Cells 1, 2, 3 and 9). The temperatures remained fairly constant throughout the monitoring period, averaging $10^{0}C$ in all the cases. Test Cells 2 and 3 with no sludge maintained acidic conditions. Buffer addition was concluded to be ineffective in controlling pH due to short circuiting of leachate.

From the study, it was concluded that addition of sewage sludge (at a rate of 0.45 kg per 115 to 160 kg of municipal solid waste) and leachate recirculation resulted in improved gas production, gas quality, and leachate quality. However, the study also encountered several problems during its operation, including an inability to accurately measure gas quantities and distribute the recirculated leachate throughout the cells, water traps in gas lines due to settlement, and problems due to freezing of pipelines.

Breitenau Landfill, Austria[46]
The Water Quality Institute of Vienna University of Technology and Waste Management Institute of Geology jointly conducted a pilot-scale research study on the reactor landfill starting in 1986.

Experimental Setup
Three test cells, 17 m (5.2 ft) deep with 2929 m² (272 ft²), 3798 m² (353 ft²), and 4622 m² (429 ft²) foot prints were constructed at the Breitenau Research Landfill, Austria, and completely filled with 35 million kg (39,000 tons), 25.6 million kg (28,000 tons), and 33.2 million kg (37,000 tons) of MSW, respectively. The cells were controlled and operated with special attention to moisture content, waste homogeneity, and leachate recirculation such that the landfill cells served as bioreactors. Test Cell 1 served as a control, Test Cell 2 received leachate recirculation, and Test Cell 3 was filled with shredded refuse and received leachate recirculation.

TABLE 3.6

Cell Composition Data, Binghamton, New York[45]

Parameter	1	2	3	Test Cell 4	5	6	9
Refuse (Mg)	8.7	7.4	6.98	6.12	6.48	7.89	7.85
Sludge (m³)	0	0	0	114	91	91	23
Water (m³)	0	114	114	0	0	0	0
Lime buffer (kg)	0	0	6800	0	6800	6800	6800
Condition	Control	Leachate recycle	Leachate recycle	Leachate recycle	Leachate recycle	Leachate recycle	Leachate recycle

TABLE 3.7

Leachate Quality Data from Day 350 to Day 600, Binghamton, New York[45]

Parameter	Test Cell						
	1	2	3	4	5	6	9
TS (%)	0.16	0.54	0.44	0.11	0.10	0.23	0.05
TVS (% of TS)	7.5	48.2	41.2	32.5	33.1	36.6	29.8
COD (mg/l)	780	1,980	1,670	180	200	80	500
TVA (mg/l)	1,120	4,310	4,310	200	270	380	160
Alk (mg/l as $CaCO_3$)	130	1,460	1,750	825	650	2,130	225
pH	6.3	5.9	6.3	6.6	6.6	6.6	6.6
Avg. gas production (m^3/d)	0.63	0.27	0.21	4.96	5.89	4.96	0.59
Avg. % CH_4	43.4	12.8	33.3	58.9	58.4	56.0	41.1
Normalized gas production (m^3/d/kg MSW × 10^{-6})	0.318	0.046	0.10	4.81	5.31	3.50	0.312
SC (mhos/cm)	660	2,930	2,160	1,510	1,850	2,890	490

TS = Total solids.
TVA = Total volatile acids.
TVA = Total volatile solids.
Alk = Alkalinity.
COD = Chemical oxygen demand.
SC = Specific conductance.

Results and Conclusions
It was observed that within a year of completion of the cells, both liquid and gaseous emissions from the cells dropped drastically. The degradation process in the landfill was observed to have followed the characteristic phases of a typical anaerobic reactor, namely, hydrolysis, acidification, methane formation, and maturation. Table 3.8 presents the average yearly leachate quality data for Test Cell 2 from 1988 to 1991 that shows a steady decrease in leachate strength. The period of decreasing leachate strength coincided with the period of steady gas production.

From the study it was concluded that under suitable operating conditions, the anaerobic degradation process could be accelerated with significant reduction in methane fermentation time. Some of the shortcomings of bioreactor operation observed were: production and escape of gas before completion of landfill, leachate ponding, and leachate toxicity due to high ammonium content.

Brogborough, United Kingdom[47]

The purpose of the Brogborough Test Cell Project was to assess the various field techniques for enhancing landfill gas production.

Experimental Setup
Six test cells were constructed (approximately 40 m (12 ft) by 25 m (7.6 ft)) adjacent to each other and separated by thick clay walls. Each cell was 20 m (6 ft) deep and filled with 15 to 20,000 metric tons (16 to 22 tons) of waste. Wastes were placed in lifts of 2 m (6.5 ft). Variables investigated were waste density and placement, waste composition, sewage sludge, leachate recirculation, and waste temperature in Cells 2 through 6, respectively. Cell 1 was the control cell. These variables were chosen for investigation because of their proven benefits and potential for incorporation into full-scale landfill sites. Cell configuration is summarized below:

- Cell 1 — control, thin layer construction (waste placed in thin layers up to 2 m (6.5 ft)

- Cell 2 — thick waste placement

- Cell 3 — leachate recirculation

- Cell 4 — air injection for temperature control, and

- Cell 5 — sewage sludge addition.

Results and Conclusions
Results indicated that a significant quantity of gas was produced within two years of initial deposition of waste. A mixture of nonhazardous industrial and commercial waste with domestic waste appeared to promote more efficient degradation based on gas production. Settlement was found

TABLE 3.8

Leachate Quality Data for Test Cell 2, Breitenau, Austria[46]

Month/Year	COD (mg/l)	BOD$_5$ (mg/l)	N(org) (mg/l)	NH3-N (mg/l)	P(total) (mg/l)	Calcium (mg/l)	pH
Nov. 1987	8200	—	200	125	2.0	1200	6.6
Mar. 1988	27000	16500	720	600	3.2	2300	6.2
Nov. 1988	4300	930	1900	1600	6.4	35	8.3
Mar. 1989	3800	770	2100	1850	9.2	30	7.9
Nov. 1989	2200	130	1300	1150	3.6	33	8.1
Mar. 1990	2650	190	1650	1550	7.2	20	7.7
Nov. 1990	1000	70	630	580	2.8	60	8.0
Mar. 1991	900	38	630	420	2.8	43	8.1
Nov. 1991	660	36	500	440	2.0	55	8.3
Feb. 1992	770	21	550	440	2.0	44	8.2

to significantly impact the integrity of the cap and gas recovery piping. Based on the monitoring data of the test cells, the following conclusions were made:

- Gas production rates were in close agreement with theoretical expectations of 5.5 to 11.0 m^3/metric t/year (170 to 350 ft^3/ton/year).

- Sewage sludge increased gas yield and quality.

- Mixed wastes enhanced gas production.

- The leachate-recirculated cell degraded organic matter faster as indicated by a significant decrease in TOC compared to other cells.

SORAB Test Cells[48]

A series of test cells, also known as Energy Loaves, were constructed on an annual basis from 1989 through 1991 at the Hagby Landfill Site in Taby, Stockholm, Sweden. The investigators used a natural digester-type reactor in order to optimize parameters controlling digestion, to minimize degradation time, evaluate gas extraction devices, investigate processing of residuals, and characterize residuals. The first Energy Loaf was 90 m (295 ft) long, 40 m (130 ft) wide, and 6 m (20 ft) high, and filled with 8200 metric tons (9000 tons) of crushed solid waste overlaid with 30 cm (1 ft) of peat for insulation. Leachate recirculation was practiced to maintain an optimum moisture content and heat the system to 35 to 40°C. Landfill gas-fueled boilers were used to heat the leachate. Gas production was reported to be an order of magnitude higher than typical. Problems were encountered with water filling vertical gas extraction wells. A later Energy Loaf employed horizontal gas collection and leachate distribution pipes.

■ FULL-SCALE LANDFILL BIOREACTOR STUDIES

In addition to laboratory-scale and pilot-scale studies of leachate recirculation and bioreactor landfill operation, a number of full-scale studies have been performed. A number of these studies are outlined below. Additional full-scale bioreactor landfill operations will be reviewed in Chapter 4 as a series of case studies.

Lycoming County, Pennsylvania[49]

Lycoming County Landfill is located 15.3 km (9.5 miles) South of Williamsport, Pennsylvania and is operated by the Lycoming County Solid Waste Department. This 53-hectare (130-acre) landfill facility serves Lycoming and other neighboring counties for a total population of 325,000. The initial fill area for Fields I, II and III was 13 hectares (31 acres) and development of additional fields has occurred when needed.

The landfill operations began in June 1978, and the site is projected to be active through 2013, based on current landfilling rates. The landfill consists of six fields numbered 1 through 6, all of them lined with PVC, the newer fields having thicker and improved liner systems. Leachate recirculation was investigated over a seven-year period.

Leachate Management Facilities
The original leachate management techniques included collection, storage, recirculation, and offsite hauling. The liner system (in addition to the site's natural features of a mantle of compacted glacial till and low permeability bed rock) consists of:

- a 30-cm (1-ft) thick sand layer containing underdrainage collection system,

- a PVC membrane liner (single 0.05-cm (20-mil) liner for Fields 1, 2 and 3; 0.076-cm (30-mil) liner for field 4 and 0.076- and 0.13-cm (30- and 50-mil) liners for field 5),

- a 15-cm (0.5-ft) sand layer containing leachate collection system piping network, and

- a 30-cm (1-ft) clay layer.

The leachate collection system consists of a series of 15-cm and 20-cm (6-in and 8-in) diameter perforated collection pipes placed in the sand layer on top of the PVC liner. The leachate collection system transports the leachate to the equalization lagoon. The PVC-lined leachate equalization lagoon with a permitted capacity of 4500 m³ (160,000 ft³) is equipped with floating aerators to keep the solids in suspension, prevent excessive odors, and provide aeration. The lagoon, from which the leachate is recirculated, also had a freeboard to handle leachate in case of emergency. Gas vents, consisting of 15-cm (6-in) diameter PVC piping in gravel-packed 1.2-m (4-ft) diameter concrete cylinders, are also provided.

Leachate Recirculation Techniques
Various techniques of leachate recirculation were tried to achieve effective distribution of leachate. Originally, it was planned to spray leachate on the operating face and other areas using spray headers. Spraying on the working face using a spray nozzle also was tried which allowed for flexibility in operation but was labor intensive and cumbersome. Spraying also caused odor problems to landfill operators and equipment. The next technique tried was to excavate small pits in the waste and fill them with leachate using a spray header. Due to the shallow depth of the landfill, the waste had limited absorption capacity and the technique was abandoned.

To increase recirculation volumes, another technique was tried incorporating trenches. Trenches were excavated on the completed sections of the landfill and filled with leachate. The absorption capacity of the trenches varied and resulted in leachate outbreaks in some parts of the landfill. Leachate outbreaks continued to occur and coincided with periods of peak infiltration and recirculation. The trench method was modified by filling the trench with auto-shredding derived waste or baled fiberglass wastes. These materials acted as wicks and transferred leachate to a larger area of the refuse thereby increasing the allowable recirculation volumes and permitting longer use of trenches. A combination of these techniques also was used. Bale-filled areas were connected to an auto waste-filled trench. An injection well also was installed in the bale-filled area using perforated concrete well rings. However, the impact of auto-shredding waste and fiberglass waste on leachate quality was not known.

Leachate Quantities

It was anticipated that due to absorption of moisture by the new waste, leachate generation would not occur until 16 to 22 years after the start of landfilling operations. However, leachate began to flow into the storage lagoon less than seven months after waste disposal began. Several factors were thought to account for the early arrival of leachate including lower waste volumes (less than design) leading to less leachate absorption, leachate channeling (resulting in inefficient absorption) and a large open area collecting precipitation. Also, the climate is humid with annual average precipitation exceeding evaporation. Within three years of landfill opening (1978), the leachate level in the equalization lagoon rose above its permitted level twice. The situations were handled by increasing recirculation quantities. Offsite hauling was started in 1982 after it was clear that leachate management by recirculation alone was not possible using the existing leachate lagoon. The lagoon also collected a significant quantity of precipitation, approximately 20,100 m^3 per year (5.3 MG/year) between January 1980 and December 1982. Consulting engineers for the facility made recommendations for the effective management of leachate which included construction of a second storage lagoon and negotiation for leachate disposal contracts with a local wastewater treatment facility.

The data on leachate quantity were from different sources including daily logs, monthly reports, planning documents, summary sheets and reports from the National Oceanic and Atmospheric (NOAA) weather station at Williamsport. The leachate quantity data included precipitation data, leachate flow measurements, lagoon level records, leachate recirculation pump records, and leachate hauling and treatment records. Although the reliability and accuracy of each source of data varies considerably, the data yield valuable information on the leachate quantities involved. The volumes of recirculation based on four sources (daily logs, monthly reports, summary sheets, and planning documents) and different pumping capacity factors were sometimes conflicting. The average recirculation rate was 24 m^3/hr (100 gpm) based on the engineer's report and 27 m^3/hr (120 gpm) based on summary sheets for the period

November 1979 through April 1981. The total quantity of leachate recirculated was around 24,600 m³ (6.5 MG) between November 1979 and January 1981. Over the first three years, over 49,200 m³ (13 MG) of leachate were recirculated. Recirculation rates approaching 3800 m³ (1 MG) per month and 19,000 m³ (5 MG) per year were recorded at the landfill. The average monthly volume of leachate hauled offsite was about 760 m³ (200,000 gal) for the period of March 1982 to June 1985.

The leachate generation quantities were estimated using a water balance method and compared with quantities derived from the lagoon balance for Field 1 for the year 1982. The water balance method estimate of 7300 m³ (2 MG) compared well with 6800 m³ (1.8 MG) derived from a lagoon balance and 8400 m³ (2.2 MG) of measured inflow into the lagoon.

It was estimated that the moisture storage capacity of the solid waste was not fully utilized and was verified by excavations that revealed dry cells which were previously considered to be at field capacity. The estimated breakdown of a water budget as of Dec. 31, 1984 is provided in Table 3.9.

TABLE 3.9

Water Budget Lycoming County[49]	
Moisture Source	Volume, m³
Percolation (1978–1984)	+121,100
Sludge water (1978–1984)	+49,200
Off-site hauling (1984–1984)	−26,500
Net utilized moisture storage capacity	143,800
Solid waste moisture storage capacity	151,400–189,300

Thus, excess or unutilized moisture storage capacity was 7,600 to 45,000 m³ (2 to 12 MG). The quantity of moisture storage capacity rendered unavailable due to clayey daily cover and cell configuration being impossible to estimate, the net available moisture storage capacity was estimated to be between zero and 38,000 m³ (0 to 10 MG) as of Dec. 31, 1984.

Leachate Quality
Samples of leachate were collected quarterly (beginning six months after landfilling commenced) and analyzed for approximately 20 parameters; 45 parameters were analyzed annually. Composite samples were taken from the lagoon through December 1979 following which grab samples were collected, typically from near the end of the leachate discharge pipe, representing raw leachate quality.

Like all leachate, the quality was highly variable, generally falling within the range reported in literature sources. The values of specific conductance, volatile acids, and manganese exceeded the upper limit of the typical ranges. The occurrence of manganese in the site soil is responsible for the high manganese content. The values of total solids, calcium, chloride and phosphate were below the lower limit of typical minimum values. The ratios of COD/TOC and BOD/COD, indicative of the age of leachate, placed the leachate in an early stage (less than five years) of landfilling when compared to typical ratios. The values of some of the key raw leachate parameters are shown in Table 3.10.

The samples were collected from each tanker whenever leachate was hauled for treatment offsite, representing leachate effluent quality. The lagoon effluent quality was less variable than raw leachate due to equalization, sedimentation, biological treatment, aeration, and dilution (due to precipitation). The lagoon effluent quality between March 1982 and January 1985 is summarized in the Table 3.11.

Gas Production
Based on the analyses of borings, samples, and mathematical modeling performed in 1983, The following was concluded:

- Field 1 was producing methane gas at a rate of 9910 m^3 (350,000 ft^3) methane/day (as of 1983), twice as much as a landfill without leachate recirculation.

- Significant gas production began in Field 1 in 1981 coincident with a sharp decline in COD and TVA and steep increase in pH.

- By 1983, more than 40 percent of methane generation capacity was exhausted in Field 1.

- Field 2 was producing methane at a rate of 10,200 m^3 (360,000 ft^3) CH_4/day (as of 1983) and was projected to increase to 22,400 m^3 (790,000 ft^3) CH_4/day in five years.

Settlement
There was no measurable settlement at the landfill. The excavations and backfilling activities, relatively shallow depth of the landfill (less than 21 m (69 ft) at the deepest point), large amount of daily cover (limiting the settlement), stockpiling of cover materials, and absence of settlement plates may have obscured settlement detection. Settlement would extend the life of the landfill because the final site development is limited by elevation and not by volume or quantity. Thus, settlement allows additional waste to be placed on completed areas.

Conclusions
Based on the performance evaluation for the Lycoming County Landfill, the following conclusions were made:

TABLE 3.10

Leachate Parameter Values, Lycoming County[49]

Parameter[a]	Number of tests	Min. (1978–1985)	Max. (1978–1985)	Avg. (1978–1985)	Avg. (1981–1985)	Range
pH	27	5.8	8.6	7.0	7.2	4.7–8.8
Alkalinity	27	404	8,300	3,100	2,400	140–9,650
BOD$_5$	26	681	28,000	7,300	5,000	4–57,700
COD	27	475	29,947	10,000	7,300	31–71,700
TOC	19	350	8,500	3,200	2,500	0–18,800
Total solids	27	1298	23,210	9,300	7,100	1,460–55,300
Volatile acids	19	223	30,730	6,600	4,000	70–27,700
Total nitrogen	10	100.3	478.5	230	140	7–1,970
Total phosphorus	10	0.03	7.2	1.0	0.46	0.2–120
Iron	27	19.5	1,095	280	230	4–2,200
Chloride	27	13	1,854	880	710	30–5,000

[a] All units in mg/l except for pH.

TABLE 3.11

Lagoon Effluent Quality, Lycoming County[49]

Parameter[a]	No. of Tests	Min.	Max.	Avg.
pH	53	6.74	8.51	7.8
Alkalinity	2	1,020	1,054	1,037
BOD₅	54	608	4,066	2,000
COD	53	1,400	5,000	3,100
Total solids	3	3,138	4,019	3,700
Total nitrogen	4	112.1	164.4	130
Total phosphates	7	BDL	2	1
Iron	31	19.08	161.5	56

[a] All units are in mg/l except for ph.
BDL — below detection limit.

- Waste degradation and methane generation were improved as a result of leachate recirculation.

- Quality of leachate stabilized more rapidly than landfills without leachate recirculation.

- Stabilization rates close to pilot-scale studies (with low recirculation rates and minimum daily cover) can be achieved.

- Clayey cover soil, high recirculation rates, and certain industrial residuals may inhibit the vertical flow of leachate resulting in incomplete use of moisture storage capacity as well as ponding within the landfill.

- The operational practices and design features for recirculation were adequate but their effectiveness could be improved.

- The major potential adverse impact of leachate recirculation involved leachate pollutant releases through irrigation drift and stormwater runoff.

- The leachate recirculation methods used were labor intensive and cumbersome although effective (injection well method being the most effective).

- Because of an inability to isolate storm water collecting on unutilized areas of the landfill from the leachate collection system, these areas generated significant quantities of leachate.

- Aerated leachate storage lagoons provided effective pretreatment of raw leachate.

- Leachate should be recirculated sufficiently to utilize moisture storage capacity rather than saturating the landfill that can lead to leachate outbreaks.

Seamer Car Landfill, United Kingdom[50]

The potential benefits of leachate recirculation including reduction of leachate volume (due to evaporation), reduction of leachate strength, rapid stabilization of wastes, and enhanced gas production were confirmed by lysimeter and pilot scale studies reported by Barber and Maris (1984). The Seamer Car Landfill investigation, initiated in 1979, was intended to determine the practicalities of leachate recirculation at full-scale landfill sites.

Landfill Description

The 2-ha site (5-acre), lined with a 0.3-cm (118-mil) HDPE, was filled with pulverized domestic waste placed at a density of 800 to 990 kg/m³ (1350 to 1680 lb/yd³). An area of one hectare (2.5 acres) was subjected to leachate recirculation by spraying, while the remaining one hectare served as the control area. Measured volumes of leachate were recirculated beginning in August 1980. Approximately 300 m³ (79,000 gal) of leachate were recirculated for five months in 1980; 3780 m³ (1 MG) and 11,400 m³ (3 MG) in 1981 and 1982, respectively, were recirculated.

Surface furrowing was found to reduce runoff and ponding problems in addition to increasing the infiltration rate significantly. The low permeability intermediate cover caused zones of saturation and lateral movement of leachate. Perched water table had developed within the recirculation area as determined by borehole investigations. All leachate was managed onsite over a three-year period. However landfill saturation eventually made offsite disposal necessary.

Conclusions

The following conclusions were made from the investigations:

- Although longer times were required than laboratory-scale studies indicated, laboratory-scale benefits could be obtained at full-scale landfills.

- Regular surface furrowing alleviated surface ponding problems.

- Intermediate cover resulted in perched water table and lateral seepage of leachate.

- Rapid reduction in leachate organic strength was achieved by increasing the waste moisture content.

- Residual COD, ammonia, and chloride concentration in leachate suggest that further treatment/dilution would be necessary before final disposal.

Delaware Solid Waste Authority[51]

The Delaware Solid Waste Authority (DSWA) operates three landfills in New Castle, Kent, and Sussex counties in Delaware. Leachate recirculation has been performed at these facilities for a number of years.

Landfill Descriptions

The Central Solid Waste Management Center (CSWMC) in Sandtown, Kent County, DE, began operations in October 1980 and has five sections, designated Areas A through E (Area E actually is two 0.4-hectare (1-acre) test cells). All of the cells are lined and equipped with leachate collection and recirculation facilities (except one of the test cells which does not have recirculation capabilities).

Leachate recirculation is applied to all the cells (excluding on of the test cells) and has been identified as one of the means (the other being landfill reclamation) of achieving the authority's objective of maximizing the reduction, reuse, recycling, and resource recovery of solid waste and minimizing landfilling. The authority refers to the term *active landfill management* to include the basic features of leachate recirculation and landfill reclamation.

At CSWMC, leachate recirculation has been accomplished by various methods including vertical recharge wells, spray irrigation systems, and surface application. Recirculation by recharge wells was found to be simplest and most affective. The recharge rate for the wells ranged from 76 to 760 l/min (20 to 200 gpm). Spray irrigation, the second preferred option, was accomplished using traveling spray irrigators with a capacity of 380 l/min (100 gpm), 30-m (100-ft) spray radius, and maximum travel distance of 210 m (680 ft). Evaporation rates of over 30 percent were measured at CSWMC.

Results

The total quantities of leachate generated and recirculated annually at CSWMC are shown in Table 3.12. The fill has been constructed in three stages in Areas A, B, and C with areas of 3.6, 7.3, and 8.1 hectares (8.8, 18, and 20 acres), respectively. At closure, trenches equipped with infiltrators are installed under the cap. The newest 8.9-ha (22-acre) cell (Area D) is double-lined including a 0.15-cm (60-mil) geosynthetic/clay composite and 0.15-cm (60-mil) HDPE liner. Leachate recirculation will be accomplished using vertical wells followed by trenches at closure. In Areas A and B, the quantity of leachate generated declined as the quantity of leachate recirculated decreased. However, the quantity of leachate treated offsite was substantial. In the case of Area C, a large portion of leachate generated was recirculated resulting in a decrease in the quantity of leachate treated offsite.

Table 3.13 shows leachate quality data for Area B over a period of ten years. Rapid decline in the organic strength of leachate, enhanced by leachate recirculation, was observed after closure in late 1988. Areas A, B, and C generated gas early during the operating period and the composition was observed to be 55 percent methane and 45 percent carbon dioxide. Unfortunately, no gas generation rates are available.

The capital cost of leachate recirculation systems (for pumping stations and piping network) constructed by DSWA ranged from $10,000 to

TABLE 3.12

Total Quantities of Leachate Generated and Recirculated Annually, Delaware Solid Waste Authority[51]			
Year	Generated, m³	Recirculated, m³	Treated, m³
1981	0	0	0
1982	2080	0	0
1983	8540	114	7600
1984	8540	117	7600
1985	5200	132	0
1986	7410	16300	0
1987	10600	10600	0
1988	12800	12800	0
1989	24200	19000	5200
1990	26800	13300	13500
1991	29500	13500	16000
1992	24600	8200	16400

$200,000 (1993 dollars) which is significantly less than the cost of a leachate treatment plant, presently being considered by DSWA, and estimated between $1,000,000 and $6,000,000. The authority found leachate recirculation to be the most economical way of handling leachate apart from the benefits of accelerated biodegradation and reduced long-term risks to the environment.

Conclusions
Based on the performance evaluation of leachate recirculating landfills, the Delaware Solid Waste Authority has demonstrated successfully that leachate recirculation results in many benefits including:

- inexpensive leachate treatment,

- accelerated biodegradation of organic portion of the waste,

- reduced long-term risk to the environment, and

- increased production of landfill gas.

German Experiences[36]
In the former Federal Republic of Germany — now unified Germany, 13 landfills were practicing leachate recirculation in 1981 using spray irrigation, spray tankers, and horizontal distribution pipes. These sites varied in size from two to 12 hectares (5 to 30 acres). For those landfills using spray irrigation, an average of 0.5 m³/ha/d (50 gal/acre/d) of excess leachate were produced; 2 m³/ha/d (200 gal/acre/d) were produced from those sites using surface percolation, and 4 to 5 m³/ha/d (400 to 500

TABLE 3.13

Leachate Quality Data for Area B, DSWA[51]

Parameter[a]	Sep. 1983	Mar. 1984	Jan. 1985	Jan. 1986	Jan. 1987	Jan. 1988	Jan. 1989	Jan. 1990	Jan. 1991	Jan. 1992	Jan. 1992
pH	5.39	7.00	5.7	5.74	5.75	6.15	6.75	6.80	7.16	7.16	7.39
COD	20,000	120	29,893	30,000	34,556	28,300	15,500	5,620	1,775	1,800	1,000
BOD	1,773	76	17,300	20,250	25,750	20,500	12,591	1,144	352	540	50
TOC	6,170	25	NA	10,000	10,000	1,900	4,950	1,178	238	540	290
TDS	NA	NA	14,800	18,600	15,999	14,713	6,558	7,726	6,497	5,100	4,900
TSS	39	19	965	137	NA	NA	1,558	502	413	170	50
Chloride	NA	NA	1,440	NA	1,500	1,683	925	1,450	650	1,100	1,200
Iron	NA	NA	972	1,050	672	1,005	596	116	104	70	12
AA[b]	NA	NA	203	NA	6,200	4,570	4,030	1,370	390	210	NA

[a] All quantities in mg/l except pH.
[b] Acetic acid.
NA — data not available.

gal/acre/d) were produced from conventional landfills without leachate recirculation. Large storage volumes (1500 to 2000 m³/ha (160,000 to 210,000 gal/acre)) were recommended.

It was observed that landfills practicing leachate recirculation since the commencement of landfill operations demonstrated a faster reduction of BOD and COD than landfills beginning leachate recirculation several years after the commencement of landfilling operation. Also, all of the landfills practicing leachate recirculation had a BOD of 1000 mg/L or less and a COD of 10,000 mg/L or less, four years after the start of landfilling operations. No increase in the concentrations of salts or heavy metals attributable to leachate recirculation was observed. Where waste was placed in thin-layers (1.8-m (6-ft) thickness), leachate was observed to have very low strength.

Bornhausen Landfill, Germany[36]

Stegman and Spendling[52] suggested that a combination of thin layer waste placement and leachate recirculation resulted in faster waste degradation and consequently faster reduction in BOD and COD concentrations in the leachate. Waste is placed in thin layers of up to 2 m (6 ft) and loosely compacted as opposed to rapid vertical filling. Thin layers promotes natural ventilation and aerobic decomposition. Penetration of oxygen into the landfill (up to 0.9 m (3 ft) depending on the density of the waste) was documented. Experiments were conducted at the Bornhausen landfill, Germany,[36] to study the thin layer process suggested by Stegman and Spendling.

Three test sites were set up with leachate recirculation. Approximately 600 m³ (21,000 ft³) waste were placed in 4-m (13-ft) layers during a period of six months. The following results were obtained:

- No increase in leachate concentration was observed after leachate recirculation was started. After 350 to 450 days, COD was less than 4000 mg/L and BOD was less than 1000 mg/L for all the test sites.

- The "thin layer" operation coupled with natural ventilation up through the drainage layer proved to be more effective than leachate recirculation.

Two other sites of approximately 0.5 hectares (1.2 acres) each at the Bornhausen landfill were constructed in 2.0-m (6.5-ft) layers, one with and the other without leachate recirculation. The time required for stabilization with leachate recirculation (230 days) was half of that required for the site without leachate recirculation (460 days).

Another significant application of leachate recirculation at the Bornhausen landfill involved the introduction of highly concentrated leachate from new landfill cells over older cells in which a stabilized leachate was already being produced. Table 3.14 presents data that demonstrate the removal of BOD and COD by two-stage leachate recirculation (leachate from the new cell recirculated over an old cell). COD and BOD reduction from 90 to 99 percent was achieved, presumably through treatment as the leachate passed through the stabilized waste.

TABLE 3.14

BOD and COD Removal by Two-Stage Leachate Recirculation, Bornhausen Landfill[36]				
	COD (mg/l)		BOD (mg/l)	
Month/Year	New	Old	New	Old
Feb. 1982	—	1473	—	60
Mar. 1982	5303	1278	1310	64
April 1982	10390	1370	5320	59
Aug. 1982	19308	1273	11970	60
Dec. 1982	4898	1083	1807	82
Feb. 1983	19385	1350	19650	61
May 1983	19675	1604	9200	173
July 1983	10780	1364	6387	91
Sept. 1983	10615	927	8750	96
Nov. 1983	21720	1271	12450	39
Dec. 1983	21470	1226	14450	37
Jan. 1984	16425	1725	6450	75

■ SUMMARY

The studies described in this chapter conclusively demonstrated the advantages of operating the landfill as a bioreactor and provided information necessary to design, construct, and operate the next generation of landfills, some of which are described in Chapter 4. Leachate and gas data from these studies are summarized and analyzed in Chapter 6. Furthermore, information derived from these and other studies provides the basis for design and operating recommendations made in Chapters 7 and 8. For convenience, major project descriptors and conclusions are provided in Table 3.15.

TABLE 3.15

Summary of Bioreactor Investigations

Location	Dimensions	Enhancement Techniques	Conclusions	Reference
Georgia Institute of Technology	4 columns: 0.9 m diameter, 3 m waste depth	Recirculation, pH control, sludge addition	• Recirculation with pH control produced low organic strength leachate faster • Sludge had no effect	2
Univ. Louisville	4 columns: 0.9 m diameter, 2.4 m waste depth	Recirculation, shredding, pH control, nutrient addition	• Recirculation with pH control produced low organic strength leachate faster • Shredding and nutrient addition no effect	35
German experiment	4 columns: 1.5 m diameter, 1.35 m waste depth	Recirculation, initial saturation, vary water input rate	• Recirculation reduced the emission of inorganic and organic pollutants • No increase in gas production or quality from enhancement	36
Newcastle Univ.	4 lysimeters: 0.5 m diameter	Recirculation, shredding, saturation vs. free draining, waste density	• Shredding increased rate of degradation • Saturation of no benefit • Lower density increased rate of waste degradation	37
Georgia Institute of Technology	10 columns: 0.9-m diameter, 3m waste depth	Recirculation, addition of priority pollutants	• Recirculation increased gas volume and rate, decreased leachate organic strength • Recirculation promoted attenuation of inorganic and organic pollutants	38

Location	Size	Treatment	Findings	Ref.
Sonoma County	5 cells: 15 m × 15 m × 3m	Recirculation, high initial water content, continuous throughput of water, septic tank pumpings addition	• Recirculation increased rate of microbial community establishment • Recirculation provided in situ leachate treatment	40
Georgia Institute of Technology	2 cells: 3 m × 3 m × 4.3 m	Recirculation, sealing of cell	• Sealed recirculation more conducive to methanogenic conditions than open air cell	42
Mountain View, California	6 cells: 10,000 m², 14 m deep	Recirculation, water addition, buffer, sludge addition	• Inconclusive regarding recirculation effects due to gas loss • Refuse analyses suggest water addition accelerates degradation	44
Binghamton, NY	9 cells: 17 m × 23 m × 6.4 m	Recirculation, sludge addition	• Recirculation and sludge addition improved gas and leachate quality	45
Breitenau landfill, Austria	3 cells: 3000–4600 m², 17 m deep	Recirculation, shredding, sludge addition	• Anaerobic digestion can be accelerated by enhancements • Concerns with ponding and increase in ammonia concentrations	46
Brogborough, UK	6 cells: 40 m × 25 m × 20 m	Thin layer construction, air injection, mixed waste, sewage sludge addition	• Large volumes of gas from cells with sludge and air addition • Liquid addition may increase gas production	47

TABLE 3.15 (Continued)

Location	Dimensions	Enhancement Techniques	Conclusions	Reference
SORAB test cells, Sweden	3 energy loaves: 95 m × 35 m × 5.5 m	Leachate heating, recirculation, shredding	• Gas volume order of magnitude greater than ordinary landfills	48
Lycoming, PA	52.6 ha, depth–max 21 m	Recirculation: • spray • trenches • injection wells	• Recirculation increases rate of waste degradation and methane generation • Ponding and saturation lead to leachate outbreaks • Injection wells most efficient	49
Seamer car landfill	2 cells, 1 ha each 4 m deep	Recirculation: spray irrigation	• Accelerated decline in leachate organic strength • Clayey intermediate cover caused ponding • Surface furrowing required	50
Delaware Solid Waste Authority	5 areas–3.6 to 8.9 ha	Recirculation: • spray • recharge wells • horizontal infiltrators	• Recirculation accelerates the biodegradation of wastes • Recirculation improved the quality of gas and leachate at low capital cost	51
Bornhauser landfill, Germany	3 cells: 50 m² × 4 m deep 2 cells: 0.6 ha, 2 m deep	Recirculation, thin layer compaction	• Recirculation cut stabilization time in half	36

4 FULL-SCALE EXPERIENCES WITH BIOREACTOR LANDFILLS—CASE STUDIES

■ INTRODUCTION

From the technical literature, telephone inquiry of state regulators, and contact with the solid waste community, a number of full-scale landfill bioreactors have been identified. These facilities evolved from demonstration projects completed in the late 1970s and early 1980s that provided essential information for the planning, design, and operation of new generation facilities. A description of many of these sites is provided below. A summary of leachate management data is provided in Table 4.1.

■ SOUTHWEST LANDFILL, ALACHUA COUNTY, FLORIDA

The active Alachua County Southwest Landfill is a 10.9-ha (27-acre) composite-lined (0.15-cm (60-mil) HDPE over 30 cm (1 ft) of clay) facility located in north central Florida. Waste first was accepted in the spring of 1988 and the facility received approximately 900 metric tons (10,000 tons) of MSW per month. Maximum landfill depth is approximately 20 m (65 ft). The landfill is permitted to recirculate up to 230 m³/d (60,000 gpd). Leachate drains by gravity through a leachate collection system to a sump and is pumped to four 340-m³ (90,000-gal) storage tanks. Excess leachate is treated using a high lime precipitation process and transported by truck to a local wastewater treatment facility.

Leachate recirculation began in September 1990 through the use of infiltration ponds (see Figs. 4.1 and 4.2). A section of the landfill was purposely not exposed to leachate recirculation to provide a comparison to the test area.[53] More than 30 million liters (8 million gal) of leachate were recycled to the landfill through the pond system from 1990 through 1992. Infiltration rates averaged between 5.3 and 7.7 l/m²/day (0.13 to 0.19 gal/ft²/day).[54]

An alternative leachate recirculation system was constructed in early 1993, providing direct injection of leachate into the landfill lifts as they were constructed (Figs. 4.3 and 4.4).[55] Horizontal pipes were placed in 2.4-m (8-ft) wide and 120 to 210-m (400 to 700-ft) long trenches filled with tire chips. The first trench was 6 m (20 ft) above the liner with subsequent trenches added at vertical intervals of 6 m (20 ft) and horizontal intervals

Portions of this chapter were reprinted from *Waste Management and Research*, Vol. 14, Debra R. Reinhart, "Full-scale Experiences with Leachate Recirculating Landfills: Case Studies," pp. 347–365, 1996, by permission of the publisher Academic Press Ltd., London.

TABLE 4.1

Full-Scale Leachate Recirculating Landfill Water Balance Data[a]

Site	Leachate Production, m³/ha/d (gal/acre/d)	Leachate Recirculation, m³/ha/d (gal/acre/d)	External Storage, m³/ha (gal/acre)	Off Site Treatment, m³/ha/d (gal/acre/d)	Design Area, ha (acres)	Active Area, ha (acres)
Alachua County	7.8 (837)	4.3 (4602)	124 (13300)	4.3 (460)	11 (27)	11 (27)
Worcester County	2.6 (275)	2.1 (230)	220 (23500)	0.64 (68)	6.9 (17)	6.9 (17)
Winfield County	19 (2000)	14 (1500)	67 (7100)	0.55 (59)	2.8 (7)	2.8 (7)
Pecan Row	2.7 (290)	1.1 (120)	690 (73600)	0	16 (40)	4.5 (11)
Lower Mt. Washington Valley	15 (1600)	9.5 (1000)	12 (1250)	4.2 (450)	3.2 (8)	0.45 (1.1)
CRSWMA	17 (1800)	12 (1200)	1600 (171000)	0	8.9 (22)	5.7 (14)
Lemons Mill Seat	2.2 (240)[b]	5 (540)	110 (11600)	NA	30 (75)	NA
Test Cell 2	2.8 (300)[b]	6.8 (720)	35 (3700)	NA	2.8 (6.9)	NA
Test Cell 3	2.8 (300)[b]	5.2 (560)	41 (4300)	NA	2.2 (5.4)	NA

[a] Based on current operational area.
[b] Estimated Using the Hydrologic Evaluation of Landfill Performance Model, excludes recirculated flow.

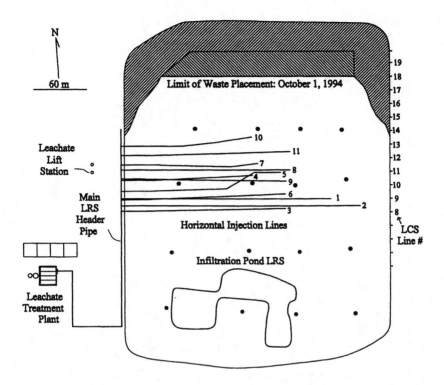

Figure 4.1 Leachate recirculation system, Alachua County, Florida, Southwest Landfill.

Figure 4.2 Leachate recirculation ponds, Alachua County, Florida, Southwest Landfill.

of 15 m (50 ft) for a total of 17 laterals. Each lateral was valved separately to allow rotation of leachate introduction. Leachate first was introduced to the injection system in February 1993. Just over 7.6 million liters (2 million gal) were pumped to the first two laterals over a period of six weeks (310 to 620 l/day per m of trench [25 to 50 gpd/ft]) at a rate of 230 to 380 lpm

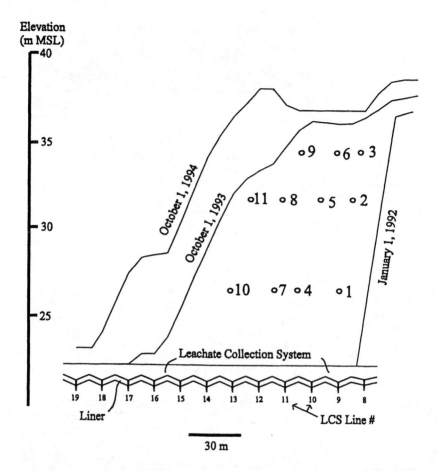

Figure 4.3 Horizontal injection system (HIL), Alachua County, Florida, Southwest Landfill.

(60 to 100 gpm) without experiencing pump discharge pressure exceeding 55 kPa (8 psi). Unlike the ponds, a direct impact on leachate quality and quantity was observed following continuous pumping to the trenches at the initial high rates. From March through September 1993, 760 to 3000 m³/month (200,000 to 780,000 gal/month) were introduced to individual laterals, and no impact on leachate quality was noted. Recirculation laterals were connected to the landfill gas recovery system in early 1994 to permit extraction of gas during the active landfill phase.[56]

■ CENTRAL FACILITY LANDFILL, WORCESTER COUNTY, MARYLAND

The Central Facility Landfill, located in Worcester County, Maryland, was constructed in the late 1980s and began operation in 1990. Initially, the

Figure 4.4 Installation of horizontal injection system (HIL), Alachua County, Florida, Southwest Landfill.

first of four 6.9-ha (17-acre) cells was constructed (Fig. 4.5). Maximum fill height was estimated to be 27 m (90 ft). Waste receipt averages 180 TPD (200 tpd). The cell is lined with a 0.15-cm (60-mil) HDPE geomembrane installed on top of natural clay soil. Leachate drains through pea gravel to 15-cm (six-inch) perforated PVC pipes that carry the leachate to sumps located at the four corners of the cell. Leachate is pumped to a 1500-m³ (400,000 gal) steel storage tank.

Leachate recirculation is accomplished using vertical discharge wells constructed using 1.2-m (four-ft) diameter perforated concrete manhole sections as seen in Fig. 4.6. The first 2.4-m (eight-ft) section rests on a concrete base and is filled with concrete to prevent short circuiting of leachate. Subsequent sections are added at each waste lift, then filled with gravel. A 5-cm (two-in) PVC standpipe is installed within each well to vent gas and permit monitoring of water depth. A schematic of the vertical well used at the Central Facility Landfill is provided in Fig. 4.7. Each well serves a 0.8-ha (two-acre) area. Leachate is pumped to the fill using flexible fire hose that can be dragged to the wells. Surface ponds are also permitted by the state to reintroduce leachate. Usually these ponds are constructed around the wells and isolated by berms.

Excess leachate is transported by truck to a local wastewater treatment facility. While minimal offsite treatment has been required, the landfill operators have expressed the opinion that the wells have limited impact area and recommended modifications which would move leachate laterally away from the wells.

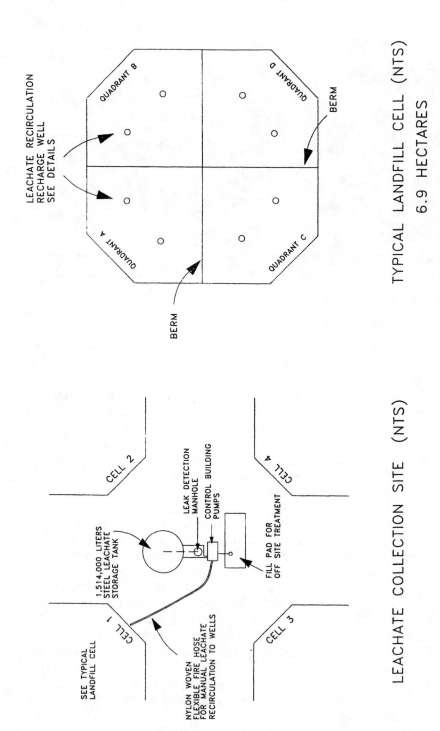

LEACHATE RECIRCULATION RECHARGE WELL SEE DETAILS

QUADRANT B

QUADRANT D

BERM

QUADRANT A

QUADRANT C

BERM

TYPICAL LANDFILL CELL (NTS)
6.9 HECTARES

CELL 2

LEAK DETECTION MANHOLE

CONTROL BUILDING PUMPS

CELL 4

1,514,000 LITERS STEEL LEACHATE STORAGE TANK

FILL PAD FOR OFF SITE TREATMENT

SEE TYPICAL LANDFILL CELL

CELL 1

CELL 3

NYLON WOVEN FLEXIBLE FIRE HOSE FOR MANUAL LEACHATE RECIRCULATION TO WELLS

LEACHATE COLLECTION SITE (NTS)

Figure 4.5 Leachate recirculation system, Worcester County, Maryland, Landfill.

Figure 4.6   Vertical leachate recharge well, Worcester County, Maryland, Landfill.

■ WINFIELD LANDFILL, COLUMBIA COUNTY, FLORIDA

The Winfield Landfill located in Columbia County, Florida, opened in September 1992. The double-liner system provided is composed of a 46-cm (18-in) drainage layer, 0.15-cm (60-mil) HDPE geomembrane, and leachate detection system installed over a 46-cm (18-in) clay soil liner, hydraulic conductivity 10^{-8} cm/sec. The cell is located above natural clay soils. The

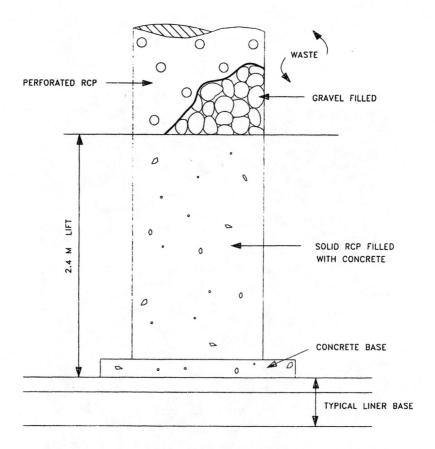

Figure 4.7 Schematic diagram of leachate recharge well, Worcester County, Maryland, Landfill.

cell slopes to the southwest to convey leachate to a single corner sump. The cell area as of spring 1995 was 2.8 ha (seven acres) with plans for an ultimate footprint of 8.9 ha (22 acres) in four expansion steps. Total depth was planned for 16 m (54 ft) providing 30 to 40 years of disposal capacity. Waste receipt averages approximately 49 TPD (120 tpd).

Collected leachate is pumped from the cell to a 190-m³ (50,000-gal) aerated lagoon from which it is either recirculated back to the landfill or tanked offsite for treatment at a local wastewater treatment facility. The facility is permitted to recirculate by surface ponds or spray, provided spraying is limited to a two-week duration at any one location. In practice, use of surface ponds has been discouraged by regulators. Recirculation is normally accomplished by pumping through one of two permanent headers through PVC lines to sprinkler heads. The spray configuration easily can be dismantled and moved. Spraying is accomplished in areas well separated from operations to an area of approximately 750 m² (8000 ft²).

Problems with ponding on the low permeability intermediate cover were reported. Ultimately, tire chips were incorporated into the intermediate cover to promote drainage. The use of low permeable cover on the slopes has reduced side seeps experienced during early operating periods.

Past operation of the landfill permitted all rainfall falling on the seven-acre site to enter the leachate collection system, producing excessive amounts of leachate. In order to minimize leachate production, the intermediate cover was sloped to the northeast to allow pumping of uncontaminated stormwater to the stormwater collection system. Leachate is recirculated primarily to the relatively flat slopes at the top of the fill; these areas are also bermed to avoid contamination of stormwater. Leachate is sprayed at lower rates on the side slopes to provide irrigation water. Operators report more than 50 percent reduction in leachate volume when recirculation is practiced.

■ PECAN ROW LANDFILL, LOWNDES COUNTY, GEORGIA

The Pecan Row Landfill is located in south Georgia near Valdosta. The landfill is located on a 39-ha (97-acre) site with an ultimate fill area of 16 ha (40 acres). Individual 1.5 to 1.6-ha (3.5 to 4-acre) cells are constructed approximately every seven months. A site plan is provided as Fig. 4.8. Maximum waste depth is planned for around 18 m (60 ft). The first phase was constructed in 1992 and became operational late the same year. Incoming waste averages around 540 TPD (600 tpd). The liner system is composed of 0.9 m (three ft) of onsite recompacted clay overlain by a 0.15-cm (60-mil) HDPE geomembrane, geonet, geotextile, and finally 0.6 m (two ft) of drainage sand. Leachate collection pipes convey leachate to a shallow, double-lined 3100- m^3 (821,000-gal) lagoon.

Leachate recirculation is accomplished by pumping (at 1500 lpm (400 gpm)) through three 15- cm (six-in) polyethylene force mains to the fill area. The recirculation system is shown schematically in Fig. 4.9. Corrugated, perforated lateral pipes branch off the force mains at 45 degree angles at 30-m (100-ft) intervals. The pipes were placed in 0.9 to 1.2-m (three to four-ft) deep, 0.9-m (three-ft) wide gravel-filled trenches dug into the waste with a backhoe. A separate recirculation system is provided at each waste lift. A total of 460 m (1500 ft) of pipe was installed in 4.5 ha (11 acres) of landfill area.

Leachate normally is pumped for one hour, then discontinued for an hour. By spring 1994 all leachate generated had been stored or recirculated as a result of high rates of evaporation from the large lagoon. To minimize leachate generation, 0.03-cm (12-mil) polyethylene sheeting is used as temporary cover over areas not receiving waste. In addition, the temporary cover is used over fill areas as daily cover. Weekly soil cover is required by the state. The cover material is removed prior to placing waste to maintain good moisture routing through the waste. Should excess leachate

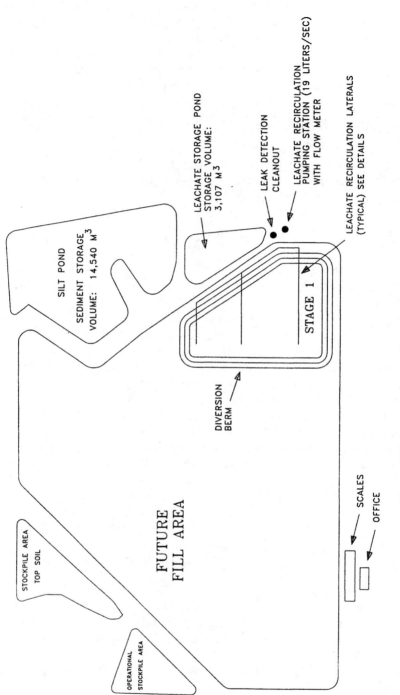

Figure 4.8 Pecan Row Landfill site plan (Valdosta, GA).

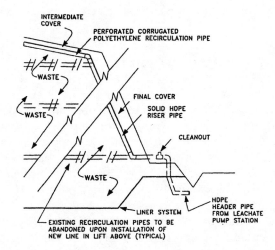

LEACHATE RECIRCULATION PIPE DETAIL (NTS)

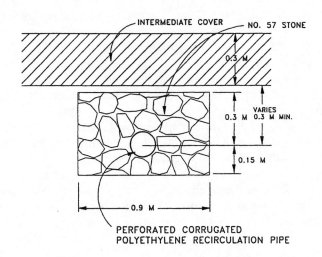

RECIRCULATION LINE DETAIL (NTS)

Figure 4.9 Pecan Row Landfill (Valdosta, GA) recirculation line detail.

accumulate, trucking to a nearby wastewater treatment facility was planned.

It was observed that once the waste was thoroughly wetted, an immediate impact on pond level occurred following each subsequent pumping. Difficulty was encountered in recirculating during early operational phases when insufficient waste was available to absorb the moisture. Also, recirculation near the waste surface or slope led to leachate outbreaks.

■ LOWER MOUNT WASHINGTON VALLEY SECURE LANDFILL, CONWAY, NEW HAMPSHIRE

The Lower Mount Washington Valley Secure Landfill, located in Conway, NH, is composed of eight hydraulically separated double-lined landfill cells (0.3 to 0.4 ha [0.75 to 1.0 acres]). Cell construction was completed in late 1991, with operations commencing in January of 1992. Waste receipt averages between 9,100 and 13,600 metric tons/year (10,000 and 15,000 tons/year). Leachate is stored in a 38-m^3 (10,000-gal) leachate collection tank.

Leachate recirculation began in May 1992 at the first of eight cells, four months after start up. The primary mode of leachate recirculation has been manual prewetting of waste using a fire hose to improve compaction and efficiently wet the waste. In addition, recirculation was accomplished using a fabricated PVC pipe manifold placed in a shallow excavation of the daily cover.

In order to minimize lateral movement of leachate, horizontal trenches were installed on waste slopes for leachate recirculation into these areas. The trenches were 1.4 to 1.8 m (four to six ft) deep, 0.9 to 1.2 m (three to four ft) wide, and 2.4 to 4.6 m (eight to 15 ft) long. However, short circuiting resulted due to the proximity of the sand drainage layer and this practice was discontinued. High leachate generation rates were experienced during the spring of 1993, which were attributed to saturation of the fill while recirculating during the previous fall and winter, followed by freezing and then the spring thaw. Consequently, leachate recirculation was temporarily discontinued in November 1993. Leachate recirculation did not result in excessive head on the liner. Efforts were made to minimize precipitation infiltration through the use of alternative daily cover and to maximize use of waste moisture holding capacity. Gas measurements suggested that leachate recirculation stimulated biodegradation of waste.

■ COASTAL REGIONAL SOLID WASTE MANAGEMENT AUTHORITY LANDFILL, CRAVEN COUNTY, NORTH CAROLINA

The Coastal Regional Solid Waste Management Authority Landfill serves three counties along the eastern coast of North Carolina, with waste receipt averaging around 320 TPD (350 tpd). The 8.9-ha (22-acre) landfill was divided into three hydraulically separated cells of approximately equal surface area. Final height is expected to be around 15 m (50 ft). A composite liner composed of 0.6 m (2 ft) of drainage sand, a fabric filter, and a 0.15-cm (60-mil) HDPE liner overlying 0.6 m (2 ft) of low permeability clay was provided. A drainage system was installed beneath the liner system to protect the liner during periods of high groundwater level. Weekly cover consists of local sandy soils while the active face is covered on a daily basis with reusable tarp. Leachate from each cell drains by gravity to manholes that are connected to a common sump. Leachate was pumped from the sump to one of two 4.5-million liter (1.2-MG) lined lagoons (4.6 m [15 ft] maximum depth). Each lagoon is equipped with two floating mechanical aerators to provide oxygen for biological treatment of leachate.

Leachate is pumped back to the first lift of waste (depth approximately 4.6 m [15 ft]) through 500 ft (150 m) of flexible hose feeding a movable vertical injection system. A steel manifold distributed leachate through twelve flexible lines to shallow black iron probes inserted into the landfill surface (Fig. 4.10). Flow to each line is controlled by individual ball

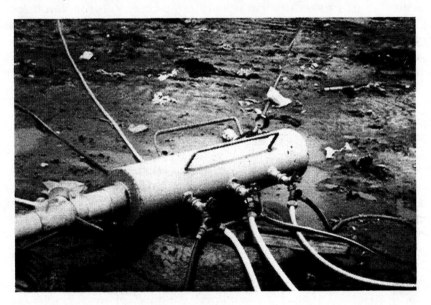

Figure 4.10 Steel manifold leachate distributed device, Coastal Regional Solid Waste Management Authority Landfill.

valves. Initially, the 1.9-cm (3/4-in) diameter probes were 1.5 m (five ft) in length with 0.32-cm (1/8-in) diameter holes drilled within 0.76 m (2.5 ft) of the bottom of the pipes. Probes were installed by driving solid pipes of similar diameter into the ground to form a hole and then inserting the probes. Early use of these probes resulted in leachate breakout at the slopes, therefore longer probes (three m [ten ft]) were fabricated to minimize breakout. The diameter was increased to 3.2 m (1.25 in) with 0.64-cm (1/4-in) diameter holes.

Leachate recirculation pump flow rates varied between 200 and 300 lpm (55 and 80 gpm) to an area approximately 30 m by 30 m (100 ft by 100 ft). Once leachate is observed at the surface near the probes, the system is moved. Generally, the system remains in any one location for two to eight days. Pressure at the recirculation manifold has been monitored and found to be around 310 kPa (45 psi).

Despite the increased depth of leachate introduction pipes, breakouts continued to occur occasionally, presumably due to shallow waste depth as well as waste saturation. In fact, leachate recirculation was temporarily discontinued due to waste saturation to the degree that heavy vehicle movement on the landfill surface was impeded. Operators attributed this problem to heavy precipitation and the inaccessibility of offsite leachate management.

Once the entire first lift is completed, horizontal trenches were to be constructed in a pattern radiating out from a central distribution box fed by the leachate recirculation pump. The horizontal system will be used until the second lift is completed at which time a new distribution system will be installed and the first system will be abandoned. This procedure will continue until the fourth and final lift is completed and the landfill is closed.

■ LEMONS LANDFILL, STODDARD COUNTY, MISSOURI

The Lemons Landfill, owned and operated by the Lemons Landfill Corporation, is located on a 66-ha (162-acre) site in Stoddard County, Missouri. Fill area at build-out is expected to be 30 ha (75 acres). Maximum depth will be 26 m (85 ft). The landfill was constructed in 1993 and began operating in October of that year. Waste is received at a rate of approximately 270 TPD (300 tpd).

A composite liner was provided, composed of a 0.15-cm (60-mil) PVC geomembrane on top of 0.6 m (2 ft) of compacted bentonite/soil, overlain with 30 cm (12 in) of pea gravel and perforated PVC piping for conveyance of leachate. Leachate is collected in two ponds that provide total storage of 3,280 m^3 (867,800 gal). Fig. 4.11 provides a schematic of the leachate management system.

Leachate recirculation is accomplished using vertical wells located at 61-m (200-ft) spacing within the fill area (Fig. 4.12). Recirculation was delayed until the area was filled and temporarily capped with 0.6 m (two ft) of clay. Recirculation began approximately one year following initial waste receipt. The leachate recirculation wells were constructed from 1.2-m (48-in) diameter precast perforated concrete pipe filled with five to 10-cm (two to four-in) diameter stone. Within the structure 30-cm (12-in) bentonite caps separate the wells into three sections. PVC pipes (10-cm (four-in) diameter) were inserted into the wells reaching each of the three sections. The well structure rests on a 3.0-m (10-ft) diameter concrete

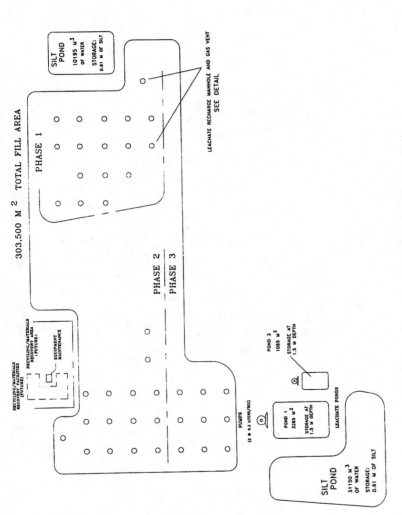

Figure 4.11 Lemon Landfill site plan (Dexter, MO).

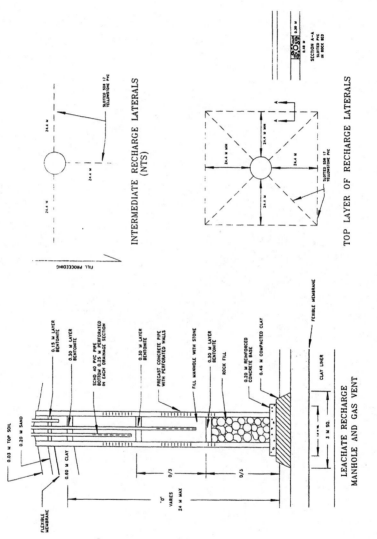

Figure 4.12 Lemon Landfill (Dexter, MO) leachate recirculation details.

pedestal underlain with a 3.6-m (12-ft) diameter, hand-tamped clay pad. Leachate distribution was supplemented by recharge laterals (7.5-cm (3–in) diameter PVC slotted pipe placed in 30-cm by 46-cm (12-in by 18–in) trenches) radiating out from each well at two depths.

After construction of the first well, planned depth of the landfill increased to approximately 30 m (100 ft). Concern over the effect of the extra well load on the liner lead designers to reconfigure subsequent wells. The new wells are installed on top of approximately 25 ft of waste using 14 ft of 42-in diameter steel casing filled with rock. Waste is placed around the casing until the pipe is partially buried. The casing is then pulled out of the waste leaving a column of rock to deliver leachate. This process continues to the full height of the landfill. Horizontal trenches are constructed as previously described and fed by three-in diameter HDPE pipes placed in the casing and surrounded by rock.

Leachate is collected in the first of two available storage ponds and recirculated back to each well at 370 lpm (100 gpm) for approximately six hours. Recirculation will continue until leachate strength is reduced significantly; after this point, leachate is diverted to the second pond and subsequently used to irrigate completed areas of the fill (capped with 0.6 cm [two ft] of clay, 0.075-cm [30-mil] PVC geomembrane, and 0.3 m [one ft] of topsoil and seeded).

■ MILL SEAT LANDFILL, MONROE COUNTY, NEW YORK

The Mill Seat Landfill, located in western New York near Rochester, hosted a bioreactor research project involving the design, construction, and monitoring of leachate recirculation in three hydraulically separated double composite-lined cells. One cell (three ha [7.4 acres]) serves as a control (gas collection only) while two test cells (2.8 ha [6.9 acres] and 2.2 ha [5.4 acres] respectively) use two different recirculation techniques. Leachate pH control was to be instituted if necessary. The test cells provide an opportunity to evaluate effects of leachate recirculation on the rate of waste stabilization, the quality of the leachate produced, and the volume of methane emitted. The three test cells comprise Stage I of the landfill that is planned to ultimately have a 38-ha (95-acre) foot print and a total waste depth of up to 34 m (110 ft).

Leachate recirculation is accomplished using two different horizontal introduction systems as shown schematically at multiple elevations in Figs. 4.13 through 4.15. The first system, installed in Test Cell 2, uses three pressurized loops constructed from 10-cm (four-in) diameter perforated HDPE pipe laid in trenches filled with crushed cullet, tire chips, or other permeable material. The loops were provided at three elevations within the cell. Collected leachate is directed to tanks providing a total of 110 m³ (30,000 gal) of storage then pumped back to the pressurized loop system.

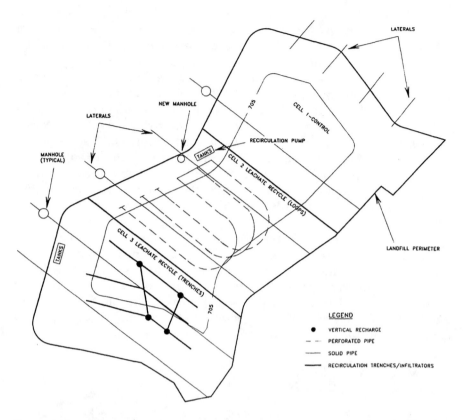

Figure 4.13   Mills Seat Landfill (Rochester, NY) recirculation layout (elevation 215 m).

The second recirculation technique, used in Test Cell 3, provides 1.3-m (4-ft) wide by three-m (10-ft) deep horizontal trenches filled with permeable wastes and installed at two elevations. Prefabricated infiltrators (Fig. 4.16) placed within the trenches enhance leachate distribution. As waste was placed on top of the trenches, chimneys were constructed to allow continued feeding of leachate to the trenches (Fig. 4.17). Leachate is introduced to the vertical chimney/wells via a tanker truck and pump. Leachate from Test Cell 3 is directed to tanks providing 76 m³ (20,000 gal) storage. Prewetting of waste on occasion using water distribution trucks also is practiced.

Recirculation rates are between 20 and 110 m³/d (5,000 and 30,000 gpd). The relative moisture content of the waste was to be monitored using gypsum blocks located *in situ* at depths of 11, 20, and 30 m (35, 65, and 95 ft) above the landfill liner. However, due to premature wetting of the blocks during waste filling, this system did not yield any data. Gas recovery ultimately will be accomplished from both the pressurized loop system and the chimneys. In addition, vertical gas wells will be installed at closure. Gas will be either flared or used to generate electricity.

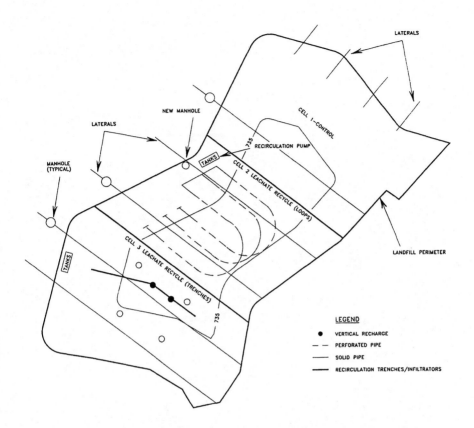

Figure 4.14   Mills Seat Landfill (Rochester, NY) recirculation layout (elevation 224 m).

■ YOLO COUNTY LANDFILL, CALIFORNIA

A demonstration initiative in Yolo County, California, funded by local and state governments was designed and constructed in 1993. Two hydraulically separated 30 m by 30 m (100 ft by 100 ft) cells were constructed to investigate impacts of leachate recirculation (Fig. 4.18). Leachate is introduced through a leach field at the top of one of the 12-m (40-ft) deep cells. The second cell serves as a control. The control cell has a single composite liner, while the recirculation cell was constructed with a double composite liner. The bottom composite layer consists of 0.6 m (2 ft) of compacted clay, a 0.15-cm (60-mil) geomembrane, a drainage net, and geotextile filter. The top liner is composed of 30 cm (one ft) of compacted clay, a 0.15-cm (60- mil) geomembrane, drainage net, and geotextile filter. Independent leachate collection and removal systems were provided.

Leachate is pumped to a distribution manifold located at the top of the test cell (Fig. 4.19). From the manifold, leachate is introduced at 25

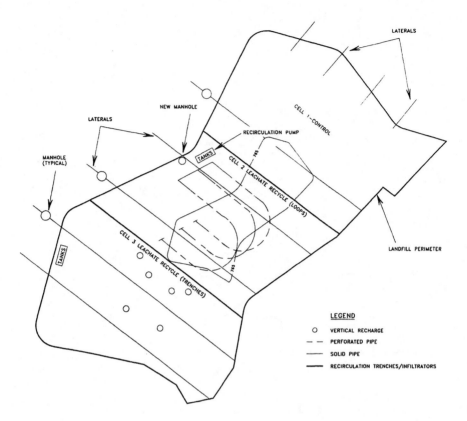

Figure 4.15 Mills Seat Landfill (Rochester, NY) recirculation layout (elevation 233 m).

locations across the surface through leachate injection pipes imbedded in a 1.5-m (five-ft) deep tire chip-filled injection pit. Warming of recirculated leachate and pH buffering were considered as additional enhancement efforts. Pressure transducers to monitor hydraulic head, temperature probes, and survey monuments to monitor settlement were provided within the fill.

■ ADDITIONAL FULL-SCALE EFFORTS

Leachate was recirculated at the Fresh Kills Landfill on Staten Island, New York between 1986 and 1993. The landfill is not lined (although it is underlain by a thick layer of low permeability natural clays) and has no leachate collection system, therefore recirculation was discontinued in late 1993. Seeps at the toe of the landfill required the installation of French drains constructed of crushed stone surrounding perforated pipe all wrapped in filter fabric. Collected leachate was pumped to the top of the landfill to a seep field dug into the waste and constructed in a similar

Figure 4.16 Horizontal infiltrators used to recirculate leachate.

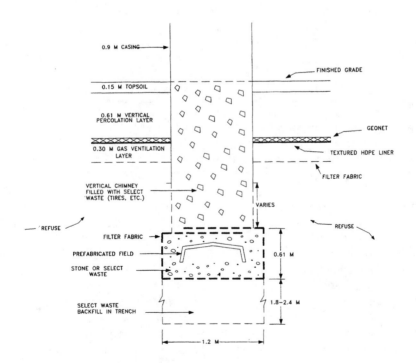

Figure 4.17 Mills Seat Landfill (Rochester, NY) prefabricated horizontal infiltration field with chimney.

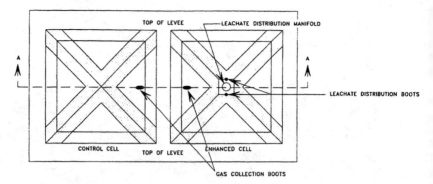

CEC CONTROL AND ENHANCED CELL COVER PLAN

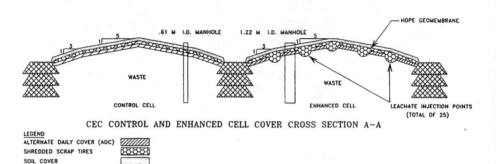

CEC CONTROL AND ENHANCED CELL COVER CROSS SECTION A−A

LEGEND
ALTERNATE DAILY COVER (ADC)
SHREDDED SCRAP TIRES
SOIL COVER
COMPACTED CLAY LEVEE

Figure 4.18 Yolo County Landfill (California) test cells.

fashion to the drain system. Each seep field served a 15- by 15-m (50 by 50-ft) area and was valved so that recirculation could be rotated from one area to another. Approximately 530 m³/d (140,000 gpd) were recirculated to an area of 20 to 34 ha (50 to 60 acres) averaging 24 m³/ha/d (2500 gal/acre/d). No clogging of the crushed stone was observed after operating for seven years.

Recirculation also has been practiced at the Gallatin National Balefill in Fairview, IL, where leachate was sprayed daily onto exposed waste surfaces using a water truck. In addition, perforated piping was installed at several elevations to provide a distribution network that doubled as a horizontal gas extraction well. The Kootenai County Fighting Creek Landfill in Idaho uses two systems for reapplying leachate. One system pumps leachate from the two aerated lagoons (total storage 4,400 m³ [1,150,000 gal]) and spray irrigated areas of the landfill that were temporarily covered and seeded to maximize evapotranspiration opportunity during summer months. Year round, leachate is pumped from collection headworks to a subsurface system composed of horizontal perforated pipes installed at

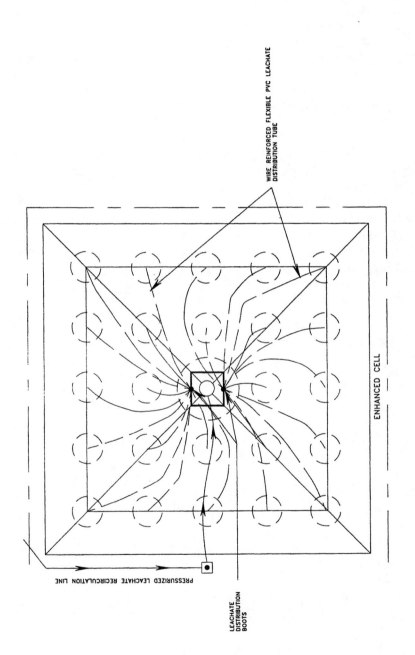

Figure 4.19 Yolo County Landfill (California) leachate distribution system layout.

30 m (100 ft) spacing under the final cover and vertical wells spaced at 91 m (300 ft).

Several full-scale bioreactor test programs took place in Sweden including a two-step degradation study conducted at Lulea, and an integrated landfill gas project involving landfills in Helsingborg, Stockholm and Malmo.[57-58] Two cells containing 450 m³ (16,00 ft³) of waste each were constructed in Viborg, Denmark, to evaluate biogas production potential and enhancement methods. One cell contained only household waste, the second household waste and yard waste with leachate recirculation provided in both.[59] Leachate recirculation was also investigated in southern Italy where experiments were conducted with distillation of leachate, recirculation of the distillate, and biological treatment of the condensate.[60] In Canada, previous concepts of dry storage are changing and leachate injection has been promoted.[61] The Rosedale Landfill in New Zealand has recirculated leachate since the mid 1980s and leachate recirculation was proposed at the Greenmount Landfill, the largest landfill in New Zealand. In San Pedro Sula, Honduras, a 250 TPD (275 tpd) landfill was constructed to recirculate leachate using a low technology methodology.[62] Leachate is introduced at the top surface of the fill and intercepted as it runs down the mound slope by pipes that convey leachate to a high permeability intermediate cover layer located at the top of waste lifts.

5

THE HYDRODYNAMICS OF LEACHATE IN RECIRCULATING LANDFILLS

■ INTRODUCTION

Implementation of leachate recirculation requires an understanding of the effects of leachate recirculation on microbially mediated processes and reactions and of the hydrodynamics of leachate flow within a landfill. Moisture distribution, effects on leachate collection systems, and optimum design and operating strategy are of particular concern to users of bioreactor landfill technology. This chapter describes the state-of-the-art understanding of leachate movement through a landfill, a brief description of mathematical modeling of leachate recirculation, and validation of model results using test cell field data.

■ LEACHATE GENERATION

Leachate Quantity

Leachate is generated primarily as a result of precipitation falling on an active landfill surface, although other contributors to leachate generation include groundwater inflow, surface water runoff, moisture from emplaced waste, and biological decomposition. The major processes controlling leachate generation and recirculation were depicted schematically in Fig. 1.1. The quantity of leachate produced is impacted by the following factors: precipitation, type of site, groundwater infiltration, surface water infiltration, waste composition and moisture content, preprocessing of waste (baling or shredding), cover design, depth of waste, climate, evaporation, evapotranspiration, gas production, and density of waste. Continuous production of leachate will occur once the absorptive capacity of waste has been satisfied. Leachate quantity is site specific and ranges from zero in arid states to nearly 100 percent of precipitation in wet climates during active landfill operation. Leachate production from new landfills occurs at relatively low rates, then increases as more waste is placed and larger areas are exposed to precipitation. Leachate production reaches a peak just before closure and then declines significantly with the provision of surface grading and interim or final cover.

Predicting Leachate Quantity

The prediction of leachate quantity has been attempted by many researchers.[10-13,63] Most commonly, leachate prediction is based on a water balance performed about the landfill by quantifying the processes shown in Fig. 1.1. The model most frequently used today is the Hydrologic Evaluation of Landfill Performance (HELP).[11] This model requires detailed information on site morphology and extensive hydrologic data to perform the water balance. The HELP computer program, which is currently in its third version, is a quasi two-dimensional hydrologic model of water movement across, into, through, and out of a landfill. Site specific information is needed for precipitation, evapotranspiration, temperature, wind speed, infiltration rates, and watershed parameters such as area, imperviousness, slope, and depression storage. The model accepts weather, soil, and design data and uses solution techniques which account for the effects of surface storage, snow melt, runoff, infiltration, evapotranspiration, vegetative growth, soil moisture storage, lateral subsurface drainage, leachate recirculation, unsaturated vertical drainage, and leakage through soil, geomembrane, or composite liners. A variety of landfill systems can be modeled, including various combinations of vegetation, cover soils, waste cells, lateral drain layers, low permeability barrier soils, and synthetic geomembrane liners. The HELP model is useful for long-term prediction of leachate quantity and comparison of various design alternatives; however, it is highly inaccurate when used to predict daily leachate production.

Hatfield and Miller[13] developed two models to better simulate leachate production at active landfills. The Deterministic Multiple Linear Reservoir Model (DMLRM) simulates the landfill moisture balance as the combined surface and subsurface discharges from three parallel reservoirs: the upper 1.2 m of the landfill, the interior of the landfill, and the surface runoff discharges. Using the conservation of volume, moisture content of cover soil and solid waste, porosity, and related parameters; total leachate volume can be estimated at any time. Validation studies showed the model to have a 5.6 percent error in prediction of cumulative discharge and an average absolute error of 56 percent in the prediction of incremental discharge volumes.

The Stochastic Multiple Linear Reservoir Model (SMLRM) was based on the same theory as the DMLRM with the exception that the three most sensitive parameters; precipitation, interception efficiency, and the maximum feasible moisture content for the waste and cover soil, were defined as probability density functions. In addition to increasing the prediction accuracy to 4.0 percent and 48 percent for cumulative and incremental discharge predictions, respectively, the use of statistical functions enables SMLRM to be applied to other landfills where less data are available.

It is safe to say that, at present, short-term prediction of leachate quantity is not an exact science and, consequently, leachate management systems must be designed to accommodate a large range of flow rates.

Internal Storage of Leachate

Internal storage of leachate within the landfill is an important concept both to the water balance used to calculate leachate generation rates and to the success of a leachate recirculation system. Internal storage of leachate is possible because the moisture content of waste as it is received is generally below the maximum absorptive capacity of the waste. Internal storage is quantified using the concept of waste field capacity, or the moisture content at which the maximum amount of water is held (through capillary forces) against gravity. The addition of more moisture will result in continuous leachate drainage.

Field Capacity

Field capacity can be determined in the laboratory by subjecting a saturated sample to 100 cm of capillary suction head and measuring the resulting moisture content. In the field it usually is calculated using a water balance approach. Moisture content is expressed on a weight basis as the weight of the water divided either by the dry or wet waste weight. On a volumetric basis, moisture content is calculated as the volume of water divided by the volume of wet waste. Field capacity is a function of the waste composition, age, density, and porosity. Table 5.1 provides the field capacity of various wastes as reported in the literature.

TABLE 5.1

Values for Field Capacity Reported in the Literature		
Field Capacity, % wet weight	Density, kg/m3 (lb/yd3)	Reference
53	213[a] (359)	63
54	500–800 (843–1350)	64
43–50[b]	500–800 (843–1350)	65
53[b]	690–950 (1160–1600)	65
47	710 (1200)	65
20–30	616[a] (1038)	66
20–35	688 (1160)	67
36.8	310 (520)	68
28.6	287 (485)	69
31–48	503 (850)	70
48	440 (735)	71
35	474 (800)	72

[a] dry
[b] shredded

Holmes[66] reported findings from an analysis of samples obtained from nineteen landfills. Field capacity was observed to decline with age due to the degradation of organic fractions that contribute to most of the absorption capacity of waste. Field capacity also decreased with increasing density due to the collapse of pore spaces available to moisture migration and retention. Other researchers have observed a significant decrease in moisture retention in baled waste.[74]

Fungaroli and Steiner[72] developed a relationship between field capacity and density as shown in Equation 5.1.

$$\theta_{fc} = 0.2 \ln \rho - 1.2 \tag{5.1}$$

where: θ_{fc} = moisture content at field capacity, and
            ρ = density, lbs/yd^3.

Field capacity also will decline with landfill depth due to the compression of lower waste layers by the waste overburden. Tchobanoglous et al.[4] reported the relationship between field capacity and waste overburden weight shown in Equation 5.2.

$$\theta_{fc} = 0.6 - 0.55\left(\frac{W}{10,000 + W}\right) \tag{5.2}$$

where: W = overburden weigh at the mid height of the waste, lb.

Fungaroli and Steiner also found that as mean particle size decreases, field capacity increases. Hentrich et al.[74] reported that shredding of waste increases the moisture holding capacity of waste.

In situ *Storage of Leachate*

In operations where moisture holding capacity of the waste is appropriately used and open areas are minimized, *in situ* storage of leachate may be adequate to manage infiltrating moisture, even during early phases of landfill operation. The Lower Mount Washington Valley Secure Landfill, Conway, NH, (LMWVSL) is composed of eight hydraulically separated small cells (0.3 to 0.4 ha [0.75 to 1.0 acres]). Leachate recirculation began in May 1992 at the first of eight cells. The primary mode of leachate recirculation is waste wetting during placement. Leachate recirculation is also accomplished using 3 m by 6 m (10 ft by 20 ft) rectangular infiltrators (fabricated PVC pipe manifolds) laid in trenches on the waste slopes. The trenches are 1.2 to 1.8 m (4 to 6 ft) deep, 0.9 to 1.2 m (3 to 4 ft) wide, and 2.4 to 4.6 m (8 to 15 ft) long. At the time of analysis leachate recirculation had not resulted in excessive head on liner and little offsite management of leachate had been necessary.[76]

A simplified water balance was developed to analyze monthly LMWVSL data. *In situ* storage can be expressed by Equation 5.3:

$$\text{STORAGE} = \text{INFILTRATION} + \text{RECIRCUALTION} - \text{LEACHATE GENERATION} \tag{5.3}$$

Infiltration is the amount of precipitation that enters the landfill and excludes evaporation, evapotranspiration, and runoff. A comparison of calculated cumulative *in situ* storage volume (calculated using Equation 5.3) and estimated waste moisture holding capacity (assuming field capacity to be 35 percent of waste, by mass and incoming waste moisture content to be 20 percent) is illustrated in Fig. 5.1. Required storage was consistently lower than the absorptive capacity of the waste. Internal storage appears to have been adequate to deal with infiltrating moisture once recirculation began, minimizing the need for offsite management of leachate. Leachate production at this site averaged 0.0016 m^3/m^2-day (1800 gal/acre-day).

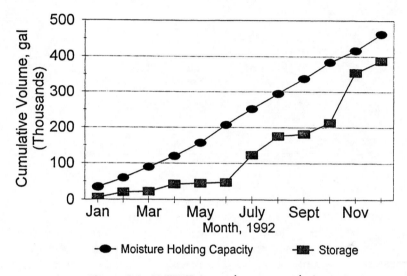

Figure 5.1 LMWVSL internal storage analysis.

In situations where large areas are open during early phases of landfill operation, excessive infiltration can lead to ponding within the landfill if sufficient *ex situ* storage and/or offsite management is not provided. This situation is illustrated by examining a landfill located in the southeastern United States. Again, *in situ* storage was calculated using Equation 5.3. In this case, evaporation data were not available; therefore, storage also includes evaporation. Site operators felt that evaporation was minimal due to excessively humid conditions during this period. As can be seen in Fig. 5.2, *in situ* storage/evaporation clearly exceeded calculated waste moisture holding capacity. Consequently, waste saturation undoubtedly occurred and surface ponding was observed. In addition, it is possible that the head on the liner exceeded permitted levels. Leachate production at this site was approximately 0.0037 m^3/m^2-day (4300 gal/acre-day).

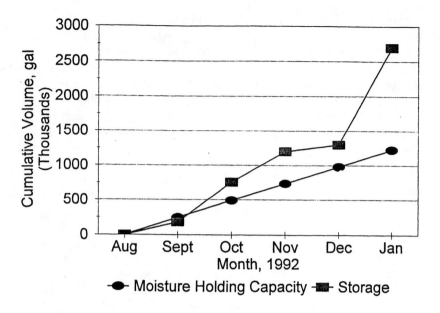

Figure 5.2   Southeastern U.S. landfill internal storage analysis.

■ MOISTURE MOVEMENT

The models described in this chapter generally assume that moisture moves through the landfill as a vertical wetting front, suggesting that leachate exits from the landfill when the internal storage capacity is exhausted, i.e., when the moisture content is at field capacity everywhere. If such a situation were true, the time of leachate arrival could be easily calculated based on the depth of the landfill and the rate of moisture infiltration. From practical operating experience it is known that leachate is frequently generated well before the time these calculations would predict. Leachate generation may occur before reaching field capacity as a result of uneven distribution of moisture, channeling, and stormwater runoff from slopes into the leachate collection system. The processes affecting moisture movement through a landfill are shown in Fig. 5.3.

Uneven moisture distribution is a natural consequence of unsaturated flow, however in landfills it is exacerbated by the heterogeneity of solid waste. Particle size of waste ranges over many orders of magnitude due to the presence of large materials such as sealed plastic bags, carpet, and plastic sheeting. Recycling probably has eliminated many of the bulky items such as appliances, batteries, and furniture. However, impermeable items and the continued use of low permeable daily and intermediate cover prevent even distribution of moisture and promote horizontal movement of leachate. In addition, gas production tends to block moisture paths in parts of the landfill during early landfill operation. As gas production declines, these flow paths reopen to leachate flow.

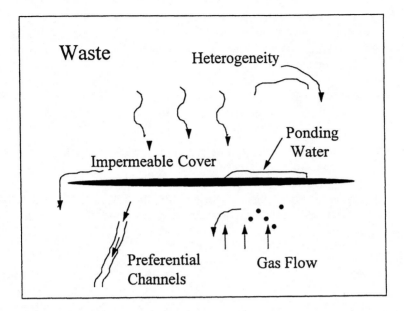

Figure 5.3 Processes affecting leachate movement through a landfill.

Channeling or fingering results in the downward movement of leachate through interconnected pores at rates faster than the uniform wetting front. The presence of preferential flow paths leads to the production of leachate far earlier than expected. Channeling declines over time as a result of landfill settlement as degradation of waste weakens the landfill structure and flow channels collapse.

■ UNSATURATED LEACHATE FLOW

Leachate movement is predominantly characterized by unsaturated flow, except for perched leachate over impermeable layers and leachate mounding near the bottom of the landfill. Darcy's Law is used to describe unsaturated flow just as it is used for saturated flow, however there are some important differences in its application. Under unsaturated conditions pressure is less than atmospheric, which explains why water will not flow into a borehole drilled into the unsaturated zone. When pressure is below atmospheric it is often known as tension or suction head and the potential is negative. This negative potential is caused by the capillary forces that hold water against gravity. Water will flow from one area that has less negative potential to another area at a more negative potential as long as the moisture content is above field capacity. The suction head at field capacity is 100 cm by definition; therefore potential is extremely negative. As the moisture content increases, the suction head declines (the potential becomes less negative) until it reaches zero at saturation.

The hydraulic conductivity in the unsaturated landfill is a function of the suction head, and therefore the moisture content. Darcy's Law, as applied to one-dimensional downward vertical flow in the unsaturated zone, is expressed as follows:

$$q = -K(\theta) - D(\theta)\frac{\partial\theta}{\partial z}$$

(5.4)

where: q = volumetric flow rate per unit surface area, L/T,
           K = hydraulic conductivity, L/T, and
           D = diffusion coefficient, L²/T.

To describe moisture distribution within the landfill, Darcy's Law is combined with the continuity equation (assuming no internal source or sink of moisture):

$$\frac{\partial\theta}{\partial t} + \frac{\partial q}{\partial t} = 0$$

(5.5)

The resulting equation is known as Richard's Equation:

$$\frac{\partial\theta}{\partial t} + \frac{\partial K(\theta)}{\partial z} - \frac{\partial}{\partial z}\left[D(\theta)\frac{\partial\theta}{\partial z}\right] = 0$$

(5.6)

The third term of Equation 5.6 describes the effects of capillary forces at the air–water interface within the pores. This term is often ignored in landfills because of the large pore structure that inhibits formation of a large suction head.[67] The resulting equation describes the kinematic theory of unsaturated flow, Equation 5.7.

$$\frac{\partial\theta}{\partial t} = \frac{\partial K(\theta)}{\partial z}$$

(5.7)

The hydraulic conductivity decreases with decreasing moisture content as shown in Fig. 5.4, reaching zero at field capacity. Hydraulic conductivity increases rapidly as moisture content reaches saturation, at which point it becomes a constant. Because of the nonlinear dependence of hydraulic conductivity on moisture content, Equation 5.7 must be solved using numerical methods.

Unsaturated Flow Characteristics

Power Law Equations

Straub and Lynch[75] were the first researchers to report on the application of unsaturated flow theory to the solid waste landfill. Power law equations (Equations 5.8 and 5.9) were used to model the unsaturated characteristics.

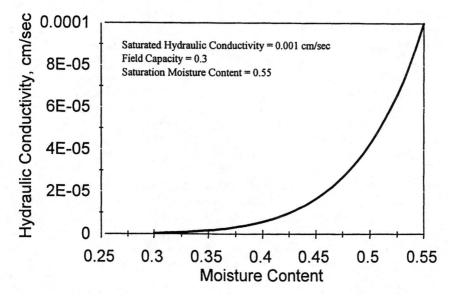

Figure 5.4 Relationship between ansaturated hydraulic conductivity and moisture content.

$$h = h_s \left[\frac{\theta}{\theta_s} \right]^{-b} \tag{5.8}$$

where: h = the suction head, L,
  h_s = saturation suction head, L,
  θ = volumetric moisture content, dimensionless,
  θ_s = saturation volumetric moisture content, dimensionless, and
  b = suction head fitting parameter.

$$K = K_s \left[\frac{\theta}{\theta_s} \right]^{B} \tag{5.9}$$

where: $K(\theta)$ = hydraulic conductivity at θ, LT^{-1},
  K_s = saturated hydraulic conductivity, LT^{-1},
  θ = volumetric moisture content, dimensionless,
  θ_s = saturation volumetric moisture content, dimensionless, and
  B = permeability fitting parameter, dimensionless.

Straub and Lynch assumed that due to the dominance of paper and fibrous materials in waste, the moisture retention characteristics of fine-grained materials could be used as a preliminary description for the moisture retention characteristics of solid waste. Values for h_s of 100 cm, b of 7, and B of 8 or 9 showed good agreement with experimental results. The saturated hydraulic conductivity was set equal to the daily average moisture application rate while the saturated moisture content was set equal to the field capacity. The setting of θ_s equal to the field capacity was justified by the theory that once the moisture content was raised above the field capacity, leachate would be produced. Field capacities ranging from 0.3 cm/cm to 0.4 cm/cm and as-placed moisture contents of 0.036 cm/cm to 0.205 cm/cm were reported. The as-placed moisture contents of 0.036 cm/cm and 0.205 cm/cm corresponded to waste with field capacities of 0.31 cm/cm and 0.375 cm/cm respectively. These wastes would then require 27 and 17 cm of moisture per meter of solid waste to reach field capacity, respectively.

Laboratory Studies of Unsaturated Flow through Waste
Korfiatis et al.[67] used a 56-cm diameter laboratory column packed with a heterogeneous mixture of approximately six-month old waste obtained from a municipal solid waste landfill to simulate the vertical movement of leachate within the landfill. The column was equipped with *in situ* pressure transducers to determine the relationship between suction head and saturation.

A 15-cm diameter column packed with waste was used to determine the saturation/suction head curve. The column was packed with waste of known moisture content. After packing was completed, pressure measurements were taken. This procedure was repeated several times at different moisture contents in order to determine the characteristics of the saturation/suction pressure curve. The power law relationships proposed by Straub and Lynch,[75] Equations 5.8 and 5.9, were used to fit the data. The saturation suction head was determined to be 6.22 cm water. Measurements of the saturated moisture content ranged from 0.5 to 0.6, a value of 0.5 was recommended. The suction head fitting parameter, b, was determined to be 1.5. However, the measurement technique did not account for channeling and most likely underestimated the suction head fitting parameter. Channeling could be accounted for by increasing the suction head fitting parameter in models that use moisture diffusion theory to model unsaturated flow. The best correlation between the equations and experiments were obtained for b equal to 4. The field capacity was found to vary from 20 to 30 percent. A value of 11 was recommended for the permeability fitting parameter, B. Saturated hydraulic conductivities ranging from 1.3×10^{-2} cm/s to 8×10^{-3} cm/s were determined for waste samples subjected to the constant head permeability test.

A sensitivity analysis was performed analyzing the importance of b and B. It was found that doubling b had little effect but that increasing B from 10 to 11 increased the cumulative volume measurement by 30 percent.

A primary difference between the Korfiatis et al. and the Straub and Lynch studies was the definition of θ_s and K_s. Korfiatis et al. defined θ_s as the actual saturated moisture content whereas Straub and Lynch defined θ_s as the field capacity of the refuse. Similarly, Korfiatis et al. defined K_s

as the measured saturated hydraulic conductivity while Straub and Lynch defined it as the moisture application rate.

Additional Methods of Characterizing Unsaturated Hydraulic Conductivity
Noble and Arnold[77] evaluated several engineering models for moisture transport within a landfill. Shredded newspaper waste was used as a solid waste surrogate in their laboratory experiments that were one-dimensional vertical flow situations. They developed the FULFILL program, a one-dimensional linearalized finite difference solution of the Richard's equation (Equation 5.6), which includes the effects of gravitational forces.

Noble and Arnold compared the power law equations (Equations 5.8 and 5.9) proposed by both Korfiatis et al.[67] and Straub and Lynch[75] to an exponential relationship:

$$K = K_s \, e^{\gamma(\theta^*-1)} \qquad (5.10a)$$

$$h = h_{max} \, e^{-a\theta} \qquad (5.10b)$$

$$\theta^* = (\theta - \theta_{ad})/(\theta_s - \theta_{ad}) \qquad (5.10c)$$

where: K = hydraulic conductivity, LT^{-1},
K_s = saturated hydraulic conductivity, LT^{-1},
h = suction head, L,
h_{max} = maximum suction head, L,
a = fitting parameter, unitless,
γ = fitting parameter, unitless,
θ^* = normalized moisture content, dimensionless,
θ = moisture content, dimensionless,
θ_s = saturated moisture content, dimensionless, and
θ_{ad} = air, dry moisture content, dimensionless.

The following data for the curve fitting constants were reported by Noble and Arnold: b equals 6 or 7, B equals 8 or 9 for the power law equations from Straub and Lynch;[75] b equals 1.5, θ_s equals 0.55, h_s equals 6.22 cm of water for small-scale studies and b = 4 and B = 11 for large-scale leaching for the power law equations from Korfiatis et al;[67] and h_{max} equals 22.4 cm of water, a equals 5 or 7, and γ equals 11 for the exponential equations from Noble and Arnold.[77] An important distinction between the exponential and power equations is that the exponential equations predict a maximum value of h_s at dry conditions (θ equals zero) whereas the power equations predict an infinite value.

Al-Yousfi[78] performed a statistical analysis based on probabilistic entropy, the concept that a system has a natural tendency to approach and maintain its most probable state, and maximization and minimization techniques ("game theory") in combination with randomness and observation techniques ("information theory") to develop Equations 5.11a and b for hydraulic conductivity as a function of saturation.

$$K(\theta) = -K_s(\theta - \theta_r) \ln\left\{1 + \left[\exp\left(\frac{-1}{\theta - \theta_r}\right) - 1\right]\frac{\theta}{\theta_s}\right\} \quad \text{for } \theta > \theta_r \qquad (5.11a)$$

$$K(\theta) = 0 \qquad\qquad\qquad\qquad\qquad\qquad \text{for } \theta < \theta_r \qquad (5.11a)$$

It was assumed that the hydraulic conductivity was zero for saturations less than the residual saturation. Equation 5.11 compares well with the equations from Noble and Arnold and Korfiatis et al. It is noteworthy that Al-Yousfi's equation was derived strictly from statistical theory and required no data.

Values of Saturated Hydraulic Conductivity in Landfills

The solution to Richard's Equation requires a determination of the saturated hydraulic conductivity. This parameter has been evaluated by many researchers. The results of their investigations are summarized in Table 5.2, where values are seen to range from 10^{-6} to 10^{-2} cm/sec. In many cases, reported hydraulic conductivities are erroneously high because of measurement techniques that include a horizontal flow component (such as pumping tests). Several researchers have found that hydraulic conductivity decreases as waste density increases.[72,81] Bleiker et al.[81] found that hydraulic conductivity decreases with depth as well, due to the impact of overburden on density. Values as low as 10^{-7} cm/sec were measured at approximately 30 m of landfill depth. This behavior contributes to saturated conditions (leachate mounding) frequently found near the

TABLE 5.2

Hydraulic Conductivity in Landfills

Reference	Hydraulic Conductivity, cm/sec	Experimental Details
72	10^{-3} to $10^{-1.7}$	Lysimeter, pumping tests, milled waste, 90–300 kg/m³
67	$10^{-2.5}$ to $10^{-2.3}$	Laboratory pump tests
79	$10^{-1.8}$ to $10^{-3.2}$	Field pump tests, 574 to 1140 kg/m³
80	$10^{-2.8}$ to $10^{-2.2}$	4-in PVC permeameter installed in test cells
68	$10^{-3.8}$ to 10^{-3}	In situ draw down tests, pump test, test pit infiltration
81	$10^{-4.4}$ to 10^{-7}	Falling head permeameter tests
54	10^{-6}	Zaslovsky's wetting-front equation applied leachate recycle ponds

bottom of the landfill. Hydraulic conductivity also declines over time as a result of chemical precipitation and biological clogging.

■ MATHEMATICAL MODELING OF LEACHATE RECIRCULATION

Because of the many parameters affecting moisture routing through a landfill, a mathematical model of the recirculating landfill was employed to more readily consider the impact of these parameters on design and operational needs. A U.S. Geological Survey (USGS) model for Saturated and Unsaturated Flow and TRAnsport, SUTRA[82,83] was used to model the recirculating landfill. SUTRA uses a two-dimensional hybrid finite element and integrated finite difference method to approximate the governing equations of flow and transport. SUTRA is capable of performing steady-state and non steady-state simulations. Modeling proceeded in two phases: steady-state modeling to perform an initial screening of SUTRA capabilities and to identify further data requirements, and transient modeling following extensive modifications to SUTRA. Two recirculation methodologies were selected for simulation, the vertical leachate infiltration well and the horizontal infiltration trench. These methodologies were selected based on their widespread use and their compatibility with the final closure of the landfill. Landfill depths of 6 and 18 meters (20 and 60 feet) were modeled.

The primary inputs to SUTRA are the physical characteristics of both the solid matrix and fluid, porosity, permeability, dispersivity, and the unsaturated flow characteristics. Porosity is input on a node-wise basis while permeability and dispersivity are input by element. The basis of the SUTRA simulation is a mesh of nodes in cartesian coordinates which are then connected to form quadrilateral elements. Output from the model provides degree of saturation (volume of water/volume of voids), fluid mass budgets, and depth of the head on the landfill liner as a function of the rate of leachate introduction and location of recirculation device(s).

The power equations (Equations 5.8 and 5.9) proposed by Korfiatis et al.[67] were used to model the unsaturated characteristics of the waste matrix. Korfiatis et al. determined the saturated suction head to be 6 cm of water for municipal solid waste. The Brooks and Corey equations with appropriate parameters were used to model the sand and gravel components of the model.[84]

$$k_r = \left(\frac{h}{h_s}\right)^{-2.75} \tag{5.12a}$$

$$\theta = \left(\frac{h}{h_s}\right)^{4} \tag{5.12b}$$

where: k_r = relative hydraulic conductivity, unitless,
  θ = volumetric moisture content, wet basis, V/V,
  h_s = saturation suction pressure, $ML^{-1} T^{-2}$, and
  h = suction pressure, $ML^{-1} T^{-2}$.

Trench Modeling

The modeling of the infiltration trench consisted of placing discrete fluid sources directly within the waste matrix. The gravel/tire chip and geotextile materials commonly used within the trench were not modeled due to dimensionality constraints and to simplify the construction of the required input files. The recirculation rates modeled were 2.0, 4.0, 6.0, and 8.0 m³/day/m of trench which bracket reported operating ranges of horizontal infiltration trenches.[55] Transient simulations of single and multiple horizontal injection lines have been conducted for a saturated hydraulic conductivity of 10^{-3} cm/s and a landfill depth of 15 m.

Vertical Infiltration Well Modeling

The modeling of the vertical infiltration well required use of a radial coordinate system and placed a series of fluid sources vertically from a level of 2 m to 13 m, discharging directly into the waste. Simulation of recirculation rates of 0.20, 0.40, and 0.80 m³/day at a hydraulic conductivity of 10^{-3} cm/s and a landfill depth of 15 m were conducted. It was also necessary to model the effect of increasing fluid pressure with depth within the well on discharge rate.

Model Results

Degree of saturation (volume of water divided by volume of pore spaces) iso-clines for the horizontal trench recirculating leachate at rates of 2.0 and 8.0 m³/day/m of trench are depicted in Figs. 5.5a and 5.5b. The influence distance was defined as the lateral distance from the trench to which the saturation had been increased above the initial condition. The degrees of saturation iso-clines suggest that flow rates of 6.0 and 8.0 m³/day/m trench result in the upward propagation of a saturated front and artesian conditions at the landfill surface.

The vertical well was modeled at flow rates of 0.20, 0.40, and 0.80 m³/day at a hydraulic conductivity of 10^{-3} cm/s for elapsed times of up to 44 days. The development of degree of saturation iso-clines for flow rates of 0.20 and 0.40 m³/day are shown in Figs. 5.6 and 5.7 respectively. It can be seen from Figs. 5.6 and 5.7 that the leachate will initially show preferential flow along the well surface. Such flow may contribute to the localized subsidence around the wells at full-scale sites. Saturations greater than 0.6 initially will develop along the well surface and slowly begin to propagate laterally and vertically as leachate attempts to percolate downward more quickly than it can be conveyed by the waste matrix.

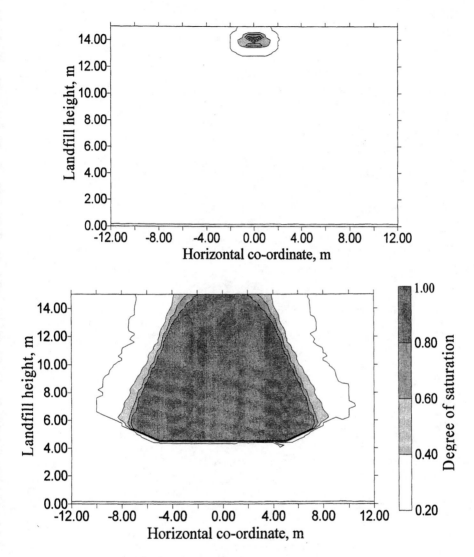

Figure 5.5 Saturation profiles for horizontal trenches recirculating 2.0 and 8.0 m^3/day/m.

■ LEACHATE RECIRCULATION FIELD TESTING

Mathematical modeling is an extremely useful tool that allows parameters to be varied without the expense of physical experimentation. However, without field verification the validity of model results is questionable. A 7.6-m deep, 3700 m^2 test cell containing some 4800 Mg of municipal solid

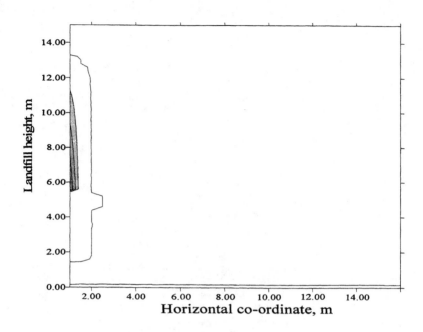

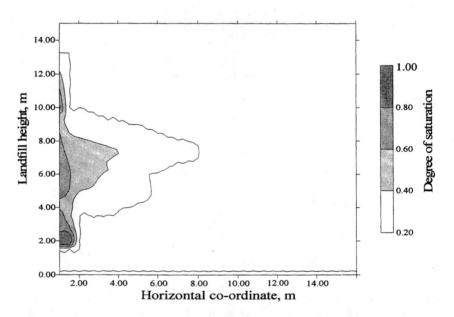

Figure 5.6   Saturation isoclines at 22 days and 44 days, vertical well recirculating 0.20 m³/day.

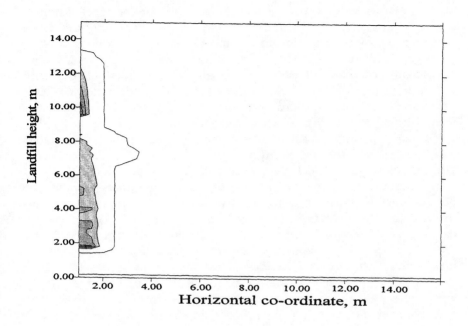

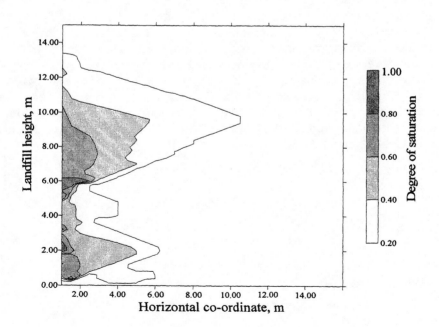

Figure 5.7 Saturation isoclines at 22 days (a) and 44 days (b), vertical well recirculating 0.40 m³/day.

waste (estimated density of 1000 kg/m^3) was constructed at the Orange County, Florida Landfill with the specific goal of monitoring leachate flow characteristics. Leachate was introduced to the cell by a multistage 1.5-hp centrifugal pump discharging to a 6-m long by 60-cm wide and 60-cm deep gravel-filled trench. Flow control was provided to permit a range of leachate flow rates. Moisture movement was examined using a dense grid of gypsum cylinders. Electrical resistance of the cylinders was measured and related to moisture content. Forty-eight cylinders were placed in horizontal lines at five levels within the cell.

A total of 49 m^3 of leachate was pumped to the test cell over thirteen weeks. Leachate was introduced at rates of 0.38 to 0.5 m^3/m/day over a 1-h period. Moisture block data were recorded on an hourly basis. The data were averaged over a week's period for graphical presentation. Where moisture blocks recorded leachate arrival, moisture content ranged from 35 to 100 percent. Moisture content iso-clines were developed for each set of weekly data. A typical iso-cline plot is presented as Fig. 5.8. From these plots, it is apparent that the wetting front propagated in a progressive fashion during periods of continuous moisture introduction. When leachate recirculation was intermittently discontinued (due to blockage of the trench) draining occurred. The pattern of moisture readings suggests that the wetting front propagated through tortuous pathways, perhaps due to the short testing period and intermittent introduction of leachate. Leachate moved approximately one m horizontally for every meter vertically downward. Horizontal movement of leachate may have been less likely to occur compared to more conventional operation due to the absence of daily cover in the test cell. The area of influence of impact was approximately 117 m^2 (maximum of 7 m from the trench) at an average infiltration rate of 2.2 × 10^{-5} cm/sec. The rate of leachate movement through the test cell was used to calculate hydraulic conductivity that ranged from 8.6 × 10^{-5} to 1.3 × 10^{-4} cm/sec for moisture contents of 40 to 70 percent, wet basis.

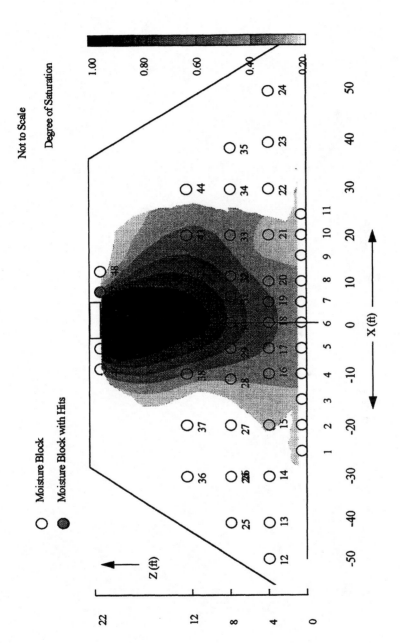

Figure 5.8 Leachate movement following introduction using a horizontal trench — test cell results.

6 THE IMPACT OF LEACHATE RECIRCULATION ON LEACHATE AND GAS CHARACTERISTICS

INTRODUCTION

Laboratory and pilot-scale studies described in previous chapters have shown that moisture control permits rapid stabilization of waste, enhances gas production, and improves leachate quality. This reduces long-term environmental consequences and liability of waste storage and improving the economics of landfills. Several dozen landfills have initiated efforts to recirculate leachate and full-scale documentation of the efficiency of this practice is now becoming possible. This chapter presents a comparison of leachate and gas characteristics at full-scale recirculating landfills with similar information from conventionally operated fills.

LEACHATE CHARACTERISTICS OF RECIRCULATION LANDFILLS

Leachate quality data were collected from five full-scale recirculating landfills located in Lycoming County, Pennsylvania; the Central Solid Waste Management Center (CSWMA), Sandtown, DE; the Southwest Landfill, Alachua County, Florida; the Central Facility Landfill, Worcester County, Maryland; and the Breitnau Research Landfill, Austria, previously described in Chapters 3 and 4. Results of preliminary analysis of the data are summarized in Tables 6.1 and 6.2. Table 6.1 provides leachate characteristics as a function of landfill stabilization phase for both conventional and recirculating landfills, while Table 6.2 compares all data. Due to differences in waste age and heterogeneity of conditions found within each landfill, explicit transitions between stabilization phases cannot be determined exactly. Nevertheless, boundaries between such phases were delineated based on the approximate magnitudes of leachate and gas strength (i.e., COD and BOD concentrations, and methane production) obtained from the records at these sites. Furthermore, a comprehensive understanding of stabilization sequence and features, as illustrated in the literature, provided the necessary guidance in dividing the stabilization histograms into their consecutive stages and projecting the overall values of leachate and gas parameters.

From these data it appears that leachate characteristics of recirculating landfills follow a pattern similar to that of conventional landfills, i.e.,

Portions of this chapter were reprinted from *Waste Management and Research*, Vol. 14, Debra R. Reinhart and A. Basel Al-Yousfi, "The Impact of Leachate Recirculation on Municipal Solid Waste Landfill Operating Characteristics," pp. 337–346, 1996, by permission of the publisher, Academic Press Ltd., London.

TABLE 6.1

Landfill Constituent Concentration Ranges as a Function of the Degree of Landfill Stabilization

Parameter	Phase II Transition		Phase III Acid Formation	
	Conventional[a]	Recirculating[b]	Conventional	Recirculating
BOD, mg/l	100–10,000	0–6893	1000–57,000	0–28,000
COD, mg/l	480–18,000	20–20,000	1500–71,000	11,600–34,550
TVA, mg/l as Acetic Acid	100–3000	200–2700	3000–18,800	1–30,730
BOD/COD	0.23–0.87	0.1–0.98	0.4–0.8	0.45–0.95
Ammonia, mg/l-N	120–125	76–125	2–1030	0–1800
pH	6.7	5.4–8.1	4.7–7.7	5.7–7.4
Conductivity, μmhos/cm	2450–3310	2200–8000	1600–17,100	10,000–18,000

moving through phases of acidogenesis, methanogensis, and maturation (although few recirculating landfills have reached maturation). These data (as summarized in Table 6.2) do not suggest that contaminants extensively concentrate in the leachate as has been promoted by critics of leachate recirculation.[86] Actually, the overall magnitude of various leachate components, during the consecutive phases of landfill stabilization, are quite comparable in both types of landfills. However, the acidogenic phase tends to be more pronounced in recirculating landfills as opposed to conventional landfills, forming a plateau with consistently high concentration of leachate constituents.[78] Such a phenomenon can be explained by the fact that uniform, high moisture contact opportunities exist in the leachate recycling landfills. On the other hand, dryness in areas of conventional landfills, accompanied by fewer chances of moisture contact and availability, minimizes the leaching opportunity in such landfills, and results in rapidly peaking leachate histograms.

Even in the case where recycled leachates are somewhat stronger than single-pass leachates, they primarily are treated inside the landfill, utilizing its storage and degradation capacity as an effective bioreactor. No extra liability and/or handling requirements will result from such cases, because leachate is repeatedly recirculated back into the landfill until its strength diminishes and stabilizes. In this respect, frequency of recirculation can be employed as a control measure to optimize landfill operations and alter leachate characteristics as desired.

Fate and Transport of Priority Pollutants

The fate of inorganic and organic compounds placed in a landfill is determined by the operative processes inherent to the natural stabilization phases occurring within the landfill. Contaminants tend to partition among

TABLE 6.1 (Continued)

	Phase IV Methane Formation		Phase V Final Maturation	
	Conventional	Recirculating	Conventional	Recirculating
	600–3400	100–10,000	4–120	100
	580–9760	1800–17,000	31–900	770–1000
	250–4000	0–3900	0	—
	0.17–0.64	0.05–0.8	0.02–0.13	1.05–0.08
	6–430	32–1850	6–430	420–580
	6.3–8.8	5.9–8.6	7.1–8.8	7.4–8.3
	2900–7700	4200–16,000	1400–4500	—

[a] Reference 16.
[b] References 46, 49, 51, 85.

TABLE 6.2

Leachate Constituents of Conventional and Recirculating Landfills — Summarizing all Phases

Parameter	Conventional[16]	Recirculating[46,49,51,85]
Iron, mg/l	20–2100	4–1095
BOD, mg/l	20–40,000	12–28,000
COD, mg/l	500–60,000	20–34,560
Ammonia, mg/l	30–3000	6–1850
Chloride, mg/l	100–5000	9–1884
Zinc, mg/l	6–370	0.1–66

aqueous, solid, and gaseous phases of the landfill. Contaminant mobility and fate is largely determined by the magnitude of the preference for one phase relative to another which is a function of the physical/chemical characteristics of both the contaminant and the phases present.

Fig. 6.1 depicts the transport/transformation phenomena that may affect the environmental fate of a landfilled contaminant. Mechanisms of mobility and transformation include biotransformation, volatilization, dissolution and advection, sorption, and chemical reactions such as precipitation, reduction, oxidation, and hydrolysis. Biotransformation and chemical reaction can reduce contaminant mass. However, a more toxic and/or mobile compound may be produced. Dissolution and advection results in the movement of the compound with the bulk flow through the refuse pore spaces. Similarly, volatilization and transport by the product gas can

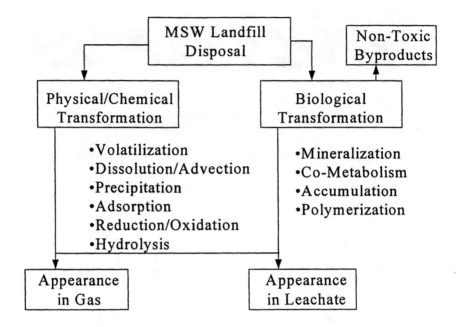

Figure 6.1 Fate and transport mechanisms for contaminants in MSW landfills.

remove the more volatile contaminants from the landfill. Sorption and precipitation can retard contaminant movement as the compound interacts with the solid phase. Transport can be influenced by compound complexation or chelation that can either retard movement if the complex becomes associated with the solid phase or enhance mobility if the compound "piggybacks" on a more soluble complexing agent.

The fate of 12 organic priority pollutants codisposed with municipal solid waste in lysimeters was investigated by researchers at the Georgia Institute of Technology.[38] The lysimeters were constructed in pairs of recirculating and single-pass operation and are described in more detail in Chapter 4. The organic priority pollutants were attenuated by abiotic and biotic transformation as well as partitioning to the waste mass. Reductive dehalogenation was the principal mechanism for halogenated compounds. The disappearance of several organic compounds suggested the possible reduction and mineralization of some aromatic compounds within the lysimeters. The conversion of organic pollutants was enhanced in the recirculating columns due to reduced oxidation-reduction potential and stimulated methanogenesis.

Relatively few metal contaminants were routinely measured in laboratory-scale landfill bioreactor studies other than iron and magnesium; however antimony, arsenic, beryllium, cadmium, chromium, copper, iron, lead, magnesium, and zinc are monitored at pilot and full-scale sites, presumably to meet regulatory requirements. Except for iron, magnesium, and, on one occasion, arsenic — at the Delaware test cells, metal concentrations were reported to be below detection limits.

At closed leachate recirculation sites, iron concentrations tended to decline with time, while remaining constant at conventionally operated landfills. Where leachate-recirculating sites were continuing to receive waste, however, iron concentrations remained elevated.

The primary removal mechanism for metals in conventionally operated landfills appears to be washout, although limited chemical precipitation may occur. In leachate-recirculating landfills, the primary removal mechanism appears to be metal sulfide and hydroxide precipitation and subsequent capture within the waste matrix by encapsulation, sorption, ion-exchange, and filtration. Gould et al.[87] found that leachate recirculation stimulated reducing conditions in lysimeters, providing for the reduction of sulfate to sulfide, which moderated leachate metals to very low concentrations. Chian and DeWalle[33] reported that the formation of metal sulfides under anaerobic conditions effectively eliminated the majority of heavy metals in leachate. In addition, metal hydroxide precipitation is enhanced under neutral or above neutral leachate conditions, which are promoted by leachate recirculation. With time, moderate to high molecular weight humic-like substances are formed from waste organic matter in a process similar to soil humification. These substances tend to form strong complexes with heavy metals. In some instances, a remobilization of precipitated metals can result from such complexation once the organic content has been stabilized and aerobic conditions begin to reestablish.[38] The potential for remobilization supports the idea of inactivating the landfill (removing all moisture) once the waste is sufficiently stabilized.

Comparison of Waste Stabilization Rates

The rate of waste placement can significantly impact leachate characteristics. Waste placed in a laboratory column will behave more or less as described in Chapter 2, following distinct phases of transition, acidogenesis, methanogenesis, and maturation. Once methanogenesis is established, the leachate organic strength generally declines. A comparison of the rate of decline in the COD for recirculating and single-pass operations was made in order to quantify the impact of leachate recirculation on leachate treatment. Because of the limited available operational information for these studies, a rigorous kinetic analysis of the sequential reactions was not possible. However, a non-linear regression of chronological, declining COD data was performed for a series of laboratory studies of leachate recirculation and conventional, single-pass operations described in Chapter 4.[2,35,38,88] From the rate of decline, COD half-lives can be calculated and compared. Half-lives are presented in Table 6.3, where it can be seen that, in most cases, recirculation accelerated leachate stabilization, as attested by the shorter half-life of COD within the recirculating lysimeters.

TABLE 6.3

COD and Chloride Half-Lives in Laboratory, Pilot and Full-Scale Leachate Recirculation Studies, Years

Study	COD		Chloride	
	Recirculating	Single-Pass	Recirculating	Single-Pass
GIT[2] (L)	0.21	0.69	0.73	0.29
GIT[88] (L)	0.19	—	1.2	—
Univ. Louisville[35] (L)	0.07	3.75	0.43	0.47
GIT[28] (L)	0.43	0.41	1.77	0.51
Austria[46] (P)	0.64	—	—	—
DSWA[51] (P)	0.32	0.27	1.8[a]	4.21
DSWA[51] (F)	1.05	—	2.89	—
Lycoming, Pa[49] (F)	0.78	—	2.58	—

L = laboratory-scale, P = pilot-scale, F - full-scale.
[a] This value is a doubling time since chloride concentration is increasing.

A similar analysis of leachate COD for full-scale leachate recirculating landfills having moved to maturation phases was made with results also provided in Table 6.3. Here, half-lives were nearly five times greater than those of laboratory recirculating lysimeters. As discussed previously, a full-scale landfill does not usually depict a single degradation phase, but rather overlapping phases representing various sections, ages, and activities within the landfill. In addition, full-scale leachate recirculating devices are generally less efficient than those used in the laboratory, therefore portions of the full-scale landfill may be relatively dry. Also, greater compaction in the field than in the laboratory will impact leachate routing negatively. Thus full-scale landfills may experience slower decay rates than laboratory fills.

Unfortunately, the relative efficacy of leachate recirculation in enhancing waste degradation relative to conventionally operated landfills at full-scale is difficult to quantify, because of the lack of conventional/recirculation parallel operations. Recognizing this limitation, leachate COD data were gathered from the literature for conventional landfills.[89] These data are plotted in Fig. 6.2. The data were analyzed to determine a COD half-life as described above. A COD half-life of approximately 10 years was calculated for conventional landfills as compared with values of 230 to 380 days for recirculating landfills.

Clearly, recirculation significantly increases the rate of the transformation of organic matter in leachate and by inference, the rate of waste

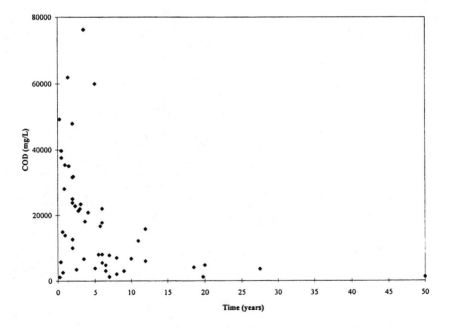

Figure 6.2 Leachate COD from conventional landfills.

stabilization. Results for conventional landfills compare favorably with values reported in the literature. Chian and Dewalle[33] reported effective landfill lives of 10 to 15 years based on gas production data and cited half-lives of 36 to 100 years from the literature. Suflita et al.[90] calculated a half-life of just over a decade for the Freshkills Landfill in New York City based on cellulose to lignin ratios.

COD is removed from single-pass landfills via washout and biological conversion and from recirculating landfills primarily by accelerated biological conversion. A conservative parameter such as chloride would be removed from single-pass landfills via washout, however in a recirculating landfill, chloride would only be removed when leachate was discharged to offsite treatment and disposal. Therefore, chloride would be expected to decline in concentration over time in a single-pass landfill, while staying relatively constant or declining more slowly in a recirculating fill (Fig. 6.3). This behavior has been observed at all scales. Chloride half-lives were calculated as described above for COD and are reported in Table 6.3. Considerably higher chloride half-lives were calculated for recirculating fills than in conventional landfills, confirming quantitatively the impact of leachate recirculation. A comparison of COD and chloride half-lives should provide an independent means of evaluating the impact of leachate recirculation on leachate components, assuming the leachate is evenly distributed and methanogenesis is occurring. The opposite effects of leachate recirculation on the conservative parameter chloride and the nonconservative parameter COD should provide a COD/chloride half-life ratio of well below one. For single-pass operations, the value should be greater than or equal to one. Limited data are available to test this hypothesis, however this trend is seen in Table 6.4.

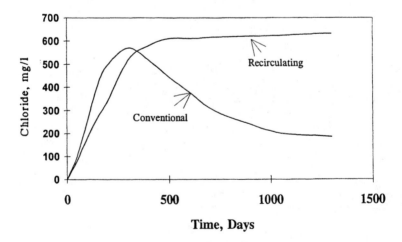

Figure 6.3 Typical behavior of chloride in conventional and recirculating landfills.

TABLE 6.4

Ratios of COD/Chloride Half-Lives		
Reference	Recirculating	Single-Pass
2 (L)	0.29	2.38
88 (L)	0.16	—
35 (L)	0.019	1.08
38 (L)	0.24	0.8
51 (P)	0	—
51 (F)	0.36	—
49 (F)	0.30	—

■ LEACHATE TREATMENT IMPLICATIONS

Ideally, the bioreactor landfill should be operated to minimize offsite management of leachate, however, eventually treatment of leachate will be necessary. Following extended recirculation, the leachate will be largely devoid of biodegradable organic matter, and will contain recalcitrant organics and inorganic compounds such as ammonia, chloride, iron, and manganese. Table 6.1 and 6.2 provide some indication of the quality of mature leachate from recirculating landfills. Treatment needs depend upon the final disposition of the leachate. Final disposal of leachate may be accomplished through codisposal at a publicly owned treatment works (POTW) or through onsite treatment and direct discharge to a receiving body of water, deep well injection, land application, or via natural or mechanical evaporation.

Leachate treatment can be challenging because of low biodegradable organic strength, irregular production rates and composition, and low phosphorous content (if biological treatment is desired). Because of the nature of leachate, physical/chemical treatment processes such as ion exchange, reverse osmosis, chemical precipitation/filtration, and carbon adsorption are the most likely options. Available leachate treatment processes are discussed by several authors.[16,33,91,92] Generally, where onsite treatment and discharge are selected, several unit processes are required to address the range of contaminants present. For example, a recently constructed leachate treatment facility at the Al Turi Landfill in Orange County, New York utilizes polymer coagulation, flocculation, and sedimentation followed by anaerobic biological treatment, two-stage aerobic biological treatment, and filtration prior to discharge to the Wallkill River.[86] Pretreatment requirements may address only specific contaminants which may create problems at the POTW. For example, high lime treatment has been practiced at the Alachua County Southwest Landfill to ensure low heavy metal loading on the receiving POTW.

■ LEACHATE QUANTITIES

Leachate volumetric data from full-scale recirculating landfill sites described in Table 5.1 and plotted in Fig. 6.4 show that greater volumes of leachate are produced as recirculation rates increase. Also, recirculated leachate represents an increasing percentage of generated flows (asymptotically approaching 100 percent) as recirculation rates increase. With the current landfill capping practices, recirculated leachate volumes will become especially dominant after landfill closure. Of the six operating sites analyzed, leachate generation rates ranged from 1.1 to 13.5 m³/ha/day (300 to 2100 gal/acre/day), with recirculated leachate representing 40 to 70 percent of leachate generated. As with conventional landfills, leachate generation is a function of climate and site characteristics, as well as leachate recirculation rates.

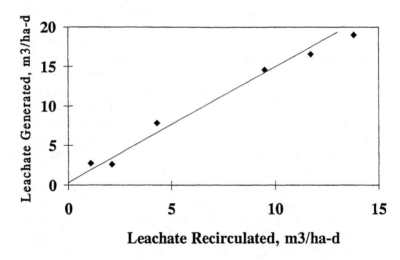

Figure 6.4   Effect of leachate recirculation on leachate generation.

Offsite disposal of leachate ranged from 0 to 59 percent of leachate generated. Data suggested that, unlike conventional landfills, the volume of leachate requiring offsite management at recirculating landfills was a function of both the volume of leachate generated and the available onsite storage. At sites where large storage volumes were provided relative to the size of the landfill cell, offsite management of leachate was minimized (frequently no offsite management was required for long periods of time). It was also observed that sites with relatively little storage were compelled to recirculate leachate at much higher rates than those with large storage volumes.

■ GAS PRODUCTION

While gas production is relatively easy to determine from laboratory lysimeters, full-scale measurement of gas emissions from active sites is more difficult to achieve. Limited data suggest that, as in lysimeters, gas production significantly is enhanced at large-scale landfills as a result of both accelerated gas production rates as well as the return of organic material in the leachate to the landfill for conversion to gas (as opposed to washout in conventional landfills). Parallel 2.5-ha (one-acre) cells operated by the Delaware Solid Waste Authority comparing conventional operation and leachate recirculation showed a ten-fold increase of gas production in the recirculating cell relative to the conventional cell.[93]

Gas emission measurements were made by researchers from the University of Central Florida in Orlando at a recirculating landfill in Alachua County, Florida using a patented device, the Flux Tube — a variation on the flux chamber used to measure surface emissions. These tests disclosed a doubling of gas production rates from waste located in wet areas of the partially recirculating landfill relative to comparably aged waste in dry areas (0.0236 std m^3/kg-yr vs. 0.0096 std m^3/kg-yr [750 std ft^3/ton-yr vs. 300 std ft^3/ton-yr]).[94] This fact was corroborated by measurements of biological methane potential (BMP) from samples obtained in wet and dry areas of the same landfill.[85] A 50 percent decrease in BMP was measured in wet samples (46 percent wet basis) over a one-year period. Negligible decreases in BMP were observed in dry samples (29 percent wet basis) over the same period.

7

LANDFILL
BIOREACTOR DESIGN

■ INTRODUCTION

In many ways, the design of the modern U.S. MSW landfill is dictated by state and federal regulation (primarily RCRA Subtitle D regulations). As discussed in Chapter 2, the design components most critical to the conventional landfill include the liner and leachate collection system, leachate management facilities, gas collection and management, and the final cap. These same components must be adapted to the landfill bioreactor to manage greater volumes of leachate, to incorporate leachate reintroduction, and to handle enhanced gas production. Critical components of the leachate recirculating landfill as described in previous chapters, include a leachate pumping station, leachate storage, leachate transmission piping, and a means for reintroduction of the leachate. This chapter addresses each of these components, providing design guidance for adapting to bioreactor operation, based on current practice. For more complete landfill design guidance the reader is referred to the many documents addressing this topic.[4-8] Since bioreactor technology is in its infancy, this chapter will reflect state-of-the-art practice; improvements in the design are expected and desired.

■ LINER/LEACHATE COLLECTION SYSTEM

The conventional liner/leachate collection system utilizes a composite, double, or double composite liner with geosynthetic and natural soil components. The landfill bioreactor requires a carefully designed liner system to accommodate extra leachate flow. At a minimum, a composite liner should be provided as encouraged by Subtitle D of RCRA. Some states, such as New York and Pennsylvania, require double composite liners irrespective of the leachate management technique employed.

The drainage system, located above the liner, is perhaps the most critical element of the collection system, and generally consists of highly permeable natural materials such as sand or gravel or a geosynthetic net. The drain must be protected by a natural soil or geosynthetic filter to minimize clogging due to particulates in the leachate as well as biological growth. Koerner and Koerner[95] concluded in a recent study that the filter should be the focus of concern in the leachate collection system because of a reduction in permeability over time. Filter clogging results from sedimentation, biological growth, chemical precipitation and/or biochemical precipitation, and is quite difficult to control. Clogging is most often experienced during the acidogenic period when organic substrates and precipitating metals such as calcium, magnesium, iron, and manganese are most highly concentrated in the leachate.[96] High substrate concentration stimulates biological growth and the development of biofilms which become encrusted with inorganic precipitates.

Since leachate recirculation promotes a shorter acidogenic phase and enhances metal removal in the waste mass, it may be assumed that recirculation would reduce the potential for clogging. Koerner and Koerner, however, expressed concern over the possibility that bioreactor operations may encourage clogging by stimulating biological processes within the cell. Koerner and Koerner suggest use of a safety factor in selecting the design filter permeability and recommend placement of a geotextile over the entire landfill foot print rather than wrapping the collection pipe. Waste with low concentrations of fines should be placed in the first layer on top of the filter.

Giroud[96] makes the following recommendations for filter selection to minimize the risk of clogging:

- sand filters and nonwoven geotextile filters should not be used,

- if a filter is used, a monofilament woven geotextile (perhaps treated with biocide) with a minimum filtration opening size of 0.5 mm and a minimum relative area of 30 percent should be selected, and

- the drainage medium should be an open-graded material, such as gravel, designed to accommodate particle and organic matter passing through the filter.

The drain should be designed to accept the excess flows expected during leachate recirculation. The depth of leachate on the liner is a function of the drainage length, liner slope, permeability of the drain and the liner, and the rate of moisture impingement. Under normal conditions, studies using the Hydrologic Evaluation of Landfill Performance Model (HELP) model have shown that the depth of the head on the liner is much less than the liner thickness and is a function primarily of the drain permeability.[97] Field experience with leachate recirculation has found excess heads on the liner on occasion, however only when *ex situ* storage and treatment is limiting (discussed further in following sections).

■ LEACHATE STORAGE

In order to gain the benefits of leachate recirculation, leachate/waste contact opportunity must be provided at rates that do not cause leachate to accumulate excessively within the landfill or emerge from landfill slopes and contaminate stormwater runoff. Proper management of leachate requires an understanding of a recirculating landfill water balance, described in more detail in Chapter 5. Precipitation falling on an active landfill will either infiltrate, run off, or evaporate. Once moisture enters the landfill, moisture holding capacity within the landfill may be sufficient to delay the appearance of leachate. Leachate generation begins when this capacity is exceeded, or, more likely, when short-circuiting occurs due to the heterogeneity in permeability within the landfill. In addition, substantial leachate flows can be generated during the active landfill phase from areas where the leachate collection system is not covered by waste and there is no opportunity for moisture absorption by the waste. In many landfills these areas are isolated using berms at the waste face and through piping and valving arrangements that allow uncontaminated water entering the leachate collection system to be diverted to stormwater management facilities.

Once filling has commenced, intermediate or final cover can be sloped so as to divert large portions of precipitation to stormwater management facilities and minimize leachate production. In some instances, plastic sheeting has been used to provide temporary cover and minimize infiltration. *Ex situ* liquid storage is vital to proper management of leachate during early phases of landfill operation, during peak storm events, and following closure of the cell, but prior to inactivation of the cell. In some areas of the country, of course, precipitation rates are so low that ensuring sufficient moisture to adequately wet the waste is more problematic than managing leachate.

Impact of Storage on Offsite Leachate Management

The impact of storage on offsite management requirements can be seen in Fig. 7.1, where data from full-scale operational sites described in Chapter 4 are plotted. As discussed in Chapter 6, at sites where large storage volumes were provided relative to the size of the landfill cell, offsite leachate management was minimized. Sites with relatively little storage recirculated leachate at higher rates than those with large storage volumes. From Fig. 7.1, storage in excess of 700 m^3/ha (75,000 gal/acre) appears to be necessary to manage leachate. Doedens and Cord-Landwehr[36] recommended storage volumes of 1500 to 2000 m^3/ha (160,000 to 210,000 gal/acre) in their investigations of German full-scale leachate recirculating landfills. The New York Department of Environmental Conservation requires storage for three month's leachate generation.

It would appear that sites with little *ex situ* storage are, in effect, using the landfill itself (the waste, drainage layers, and the leachate collection and recirculation piping) and a storage vessel. Such *in situ* storage is not necessarily detrimental if ponding is avoided and heads on the liner are controlled to meet regulatory requirements. In fact, in operations where moisture holding capacity of the waste is appropriately

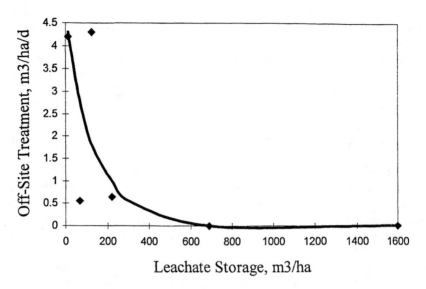

Figure 7.1   Effect of leachate storage on offsite treatment requirements.

used and open areas are minimized, as discussed in Chapter 5, *in situ* storage of leachate may be adequate to manage infiltrating moisture, even during early phases of landfill operation. Doedens and Cord-Landwehr[36] estimated that the additional storage provided from homogeneous distribution of reintroduced leachate amounted to some ten times the volume of leachate generated. However, in situations where large areas are open during early phases of landfill operation, infiltration can lead to ponding within the landfill and excessive head on the liner if *ex situ* storage and/or offsite management is not sufficient to permit timely removal of leachate.

Storage Sizes

The type of storage facility provided varies from one site to another, and includes elevated steel tanks (Fig. 7.2), lined concrete tanks, underground storage tanks, and lined ponds. Construction material must withstand the corrosive nature of leachate. Often these ponds are covered to minimize the introduction of stormwater. However, evaporation opportunity is lost under these circumstances. These facilities must be sized both to provide adequate capacity during precipitation events and to ensure the availability of sufficient volumes of leachate to recirculate at an effective rate.[98] In any case, storage should exceed 700m³/acre in wet climatic areas. When precipitation exceeds the design storm event, offsite treatment will be required and accommodations for transport and treatment of these flows must be made. It has been observed, however, that leachate storage facilities have been frequently undersized, leading to excessive head on the liner, ponding, side seeps, and costly transportation of leachate for *ex situ* treatment.

Figure 7.2 Steel above-ground leachate storage tank.

Baetz and Onysko[98] provided a more quantitative method of sizing storage facilities based on a simple hydrologic balance about a recirculating landfill. The methodology assumes the moisture absorption capability of the landfill is exhausted. Primary flows in and out of the landfill involve precipitation, recirculated flows, leakage through the liner, and leachate removed from the storage facility for *ex situ* management. They also assume that the storage facility is empty at the start of each precipitation event and full at the end of the event. They recommend setting the volume of storage at the maximum of two volumes: the volume required to recirculate when no precipitation has occurred (to compensate for leakage through the liner) and the volume of leachate generated during a peak storm event (a function of the intensity and duration of the storm and the area collecting precipitation). The first volume (V_1) can be calculated as the product of the leakage rate through the liner and the design precipitation interevention time:

$$V_1 = Q_l\, t_{ie} \qquad\qquad (7.1)$$

where: V_1 = volume for first condition, L^3
        t_{ie} = design precipitation interevent time, T, and
        Q_l = leakage rate, L^3/T.

Q_l can be determined using Darcy's Law or empirical formulae developed for leakage through composite liners.[97]

The second volume (V_2) can be calculated by determining the percolation rate, Q_P, and the design precipitation event duration.

$$Q_P = i_d P A \qquad (7.2)$$

where: Q_P = percolation rate, L^3/T
  i_d = design precipitation intensity, L/T
  P = percolation factor (the fraction of rainfall that percolates out through the base of the cover), and
  A = area, L^2

$$V_2 = Q_P t_e \qquad (7.3)$$

where: V_2 = volume for second condition, L^3, and
  t_e = duration of precipitation event, T.

The design volume is then the maximum of V_1 and V_2.

The methodology can be illustrated by the following example. A 5-ha (12 acre) hypothetical landfill is underlain by a low permeability soil liner. The percolation factor is 0.2. Design rain intensity is 100 mm/hr (4 in/h), lasting 2.8 hrs (a 25-year storm). The time between events averages 38.3 hrs. The leakage rate is estimated at 5×10^{-3} m³/sec (0.8 gpm). The percolation rate is calculated to be 3.5×10^{-3} m³/sec (56 gpm). Therefore, the volume for the first condition is 7 m³ (1,900 gal) and the second, 2,700 m³ (720,000 gal). The design volume, therefore, is driven by the storm event (as would be expected) and is 2,700 m³ (720,000 gal).

■ LEACHATE REINTRODUCTION SYSTEMS

The efficiency of leachate distribution and waste moisture absorption varies with the device used to recirculate leachate. Full-scale methods currently employed include prewetting of waste, spraying, surface ponds, vertical injection wells, and horizontal infiltration devices. These methods also differ in leachate recirculating capacity, volume reduction opportunities, and compatibility with active and closed phases of landfill operation. The advantages and disadvantages of each method are summarized in Table 7.1. Table 7.2 provides a listing of hydraulic application rates used at full-scale operating sites.

Prewetting of Waste

Prewetting of waste has been practiced for many years as a method for increasing compaction efficiency. More recently, leachate has been used as the wetting agent. Waste wetting is most commonly accomplished using water tankers[36] (Fig. 7.3) or by manual spraying using a fire hose. In addition to compaction enhancement, prewetting has advantages in terms of simplicity, evaporation opportunity, and a uniform and efficient use of waste moisture holding capacity. This technique has been rarely used in large-scale operations because of its labor-intensive nature. Obvi-

TABLE 7.1

Comparison of Frequently Used Leachate Recirculation Devices

Recirculation Method	Disadvantages	Advantages
Prewetting	• Labor intensive • Blowing of leachate • Enhances compaction (may interfere with leachate routing) • Incompatible with closure	• Simple • Uniform and efficient wetting • Promotes evaporation
Vertical injection wells	• Subsidence problems • Limited recharge area • Interference with waste placement operations	• Relatively large volumes of leachate can be recirculated • Low cost materials • Easy to construct during and following waste placement • Compatible with closure
Horizontal trenches	• Potential subsidence impact on trench integrity • Potential biofouling may limit volume • Inaccessible for remediation	• Low cost materials • Large volumes of leachate can be recirculated • Compatible with closure • Unobtrusive during landfill operation
Surface ponds	• Collect stormwater • Floating waste • Odors • Limited impact area • Incompatible with closure	• Simple construction and operation • Effective wetting directly beneath pond • Leachate storage provided
Spray irrigation	• Leachate blowing and misting • Surface precipitation leads to decreased permeability • Cannot be used in inclement weather • Incompatible with closure	• Flexible • Promotes evaporation

TABLE 7.2

Full-Scale Leachate Recirculation Hydraulic Application Rates

Recirculation Method	Application Rates	Reference/Comments
Vertical injection wells	(A) 0.23 to 0.57 m³/hr per 6.4 cm diameter well 0.07 to 0.17 m³/m² landfill area/day	(A) Noncontinuous injection rate industrial landfill, hydraulic conductivity 10^{-2} cm/sec
	(B) 4.6 to 46 m³/hr per 1.2 m diameter well 0.005 to 0.09 m³/m² landfill area/day	(B) 51 (noncontinuous injection rate municipal landfill, estimated hydraulic conductivity 10^{-4} cm/sec)
Horizontal trenches	0.31 to 0.62 m³/m of trench length/day at 14 to 23 m³/hr	55 (early in application period)
Surface ponds	0.0053 to 0.0077 m³/m²-day	54
Spray irrigation	(A) 0.73 m³/m² of landfill area/day (B) 0.001 to 0.0032 m³/m² of landfill area/day	(A) 51 (intermittent application)
		(B) 99

Figure 7.3 Prewetting of waste using a tanker truck.

ously this technique cannot be used following landfill closure, when it would be replaced by some form of subsurface injection.

Leachate Spraying

Recirculation of leachate via surface spraying has been practiced at the Seamer Carr Landfill in England,[99] and landfills in Delaware,[51] the Kootenai County Fighting Creek Landfill in Idaho,[100] and the Winfield Landfill in Florida. Problems were encountered at the Seamer Carr landfill with the development of a solid hard-pan at the surface due to chemical precipitation of leachate constituents when exposed to air. Therefore, surface furrowing was necessary to increase infiltration rates into the landfill. Leachate blowing and misting as well as odor problems were described by Watson,[51] problems that have led many state regulators to ban spraying of leachate. Doedens and Cord-Landwehr[36] recommend spraying only when COD is below 1000 mg/l and observed that flows were reduced by 75 percent when spray was utilized. Leachate spraying is quite flexible, the systems can be constructed to be easily moved from one area to another to maximize applications rate and avoid active areas. Spraying provides the greatest opportunity for volume reduction of all recirculation methods used to date. Spraying cannot be used during periods of rain or freezing conditions and is not compatible with the application of an impermeable cover at closure.

Surface Ponds

Leachate recirculation using surface infiltration ponds has been successfully accomplished at several landfills in Florida (Fig. 4.2) and with less success in Delaware. Ponds are simple to construct and operate by removing one to two meters of waste and introducing leachate. However, they consume significant portions of the active landfill that are not available for waste disposal. Ponds collect stormwater and can be the source of odors, although this has not been reported to be a problem at Florida landfills, perhaps due to the lower organic strength of Florida leachates. Unless moved frequently, ponds will have limited recharge areas, as was illustrated by recent data from the Alachua County site where waste moisture content below the pond averaged 46 percent of wet weight while moisture content in areas immediately adjacent to the pond was only 29 percent.[85] Floating waste has been a problem at the Alachua County site as well, leading to the abandonment of ponds in favor of a horizontal injection system. As with other surface introduction methods, ponds will not be compatible with an impermeable final landfill cover.

Vertical Injection Wells

Vertical wells were, at one time, the most popular engineered approach to leachate recirculation and are used at the Worcester County Landfill, in Delaware Landfills, the Lemons Landfill, and the Kootenai County Fighting Creek Landfill. Well spacing varies anywhere from one well per 0.10 ha (0.25 acre) of surface area to one per 0.8 ha (2 acres). If wells are spaced too closely, they may interfere with waste placement and compaction. Concern has also been expressed over possible tearing of the bottom liner if the well rests directly on the geomembrane, as well as problems with well integrity during landfill subsidence. Generally the bottom section of the well is not perforated to minimize leachate short circuiting. Usually the wells are installed as each lift of waste is placed by stacking sections of large diameter perforated concrete pipe (frequently manhole sections). Typical vertical wells are shown in Figs. 4.6, 4.7, and 4.12. Wells either are fed by permanent leachate transmission piping or by leachate tank trucks as shown in Fig. 7.4. Infiltration rates also appear to be enhanced if rest periods are provided between pumping events.

Horizontal Subsurface Introduction

Horizontal subsurface introduction has been used at the Alachua County Southwest Landfill, the Lower Mount Washington Valley Secure Landfill, the Fresh Kills Landfill, the Pecan Row Landfill, and the Lycoming Landfill in Pennsylvania. Horizontal infiltrators (hollow half pipes imbedded in gravel) are used in landfills in Delaware and the Mill Seat Landfill in New York. Typical horizontal devices are shown in Figs. 4.3, 4.4, and 4.9. In all cases, horizontal trenches are dug into the waste and filled with a permeable material such as automobile waste (Lycoming County), gravel (Delaware and Fresh Kills) or tire chips (Alachua County) surrounding a perforated pipe (HDPE or PVC). Construction without fill material is much quicker, however if a pipe break should occur, transmission of leachate

Figure 7.4 Influence of leachate to a vertical well using a tanker truck.

in the trenches would be significantly reduced. Infiltrators such as those shown in Fig. 4.16 often are incorporated in the trenches to direct the leachate downward. Leachate is either fed to perforated piping by gravity or injected under pressure. Horizontal systems can be used during active phases of the landfill or at closure if constructed as part of the cover system. These systems are easily constructed by landfill personnel.

Both Alachua County and New Hampshire sites reported that overuse of trenches led to significant increases in leachate collection rates as well as concentration spikes. Landfill subsidence adversely may affect the integrity of horizontal systems although no evidence of this problem has been found to date. Critical areas such as connections at transmission manifolds should be constructed with flexible piping and valves should be well supported to accommodate settlement. Breaks within the trenches are not as critical because the trench fill material will continue to transmit leachate.

Large quantities of leachate can be successfully introduced to trenches, although long-term use may result in biofouling of trench fill materials and a consequential reduction in permeability. Flow rates and pump pressure were monitored at the Alachua County Southwest Landfill (described in Chapter 4) during leachate recirculation using horizontal injection lines.[55] Increases in pressure and declining flow rates were observed over a 19-month period due to blockages. However substantial amounts of leachate were transmitted despite this effect.

■ LEACHATE RECIRCULATION SYSTEM DESIGN

Based on evaluation of available information the most practical and efficient recirculation methodology uses horizontal devices, vertical devices, or a combination of horizontal and vertical systems. Design criterion for placement of reintroduction devices is scarce and typically based on prior experience. Other issues remain uncertain in designing for full-scale leachate recirculation including the determination of the area of influence of recirculation devices, the effect of leachate recirculation on leachate collection systems, and appropriate recirculation flow rates. As described in Chapter 5, a U.S. Geological Survey software package entitled SUTRA (Saturated and Unsaturated Transport Model), a finite element simulation model for saturated/unsaturated flow, was modified to model the hydrodynamics of leachate flow through a landfill following introduction of leachate using vertical and horizontal introduction devices. This program generated isoclines of pressure and saturation data that have been used to develop design guidance.

Horizontal Trenches

The results of the modeling efforts for the horizontal trench are presented in Fig. 7.5 where the distance from the trench reached by reintroduced leachate as a function of the flow rate per unit length of trench is plotted. As can be seen in Fig. 7.5, the impact distance increases as flow rate

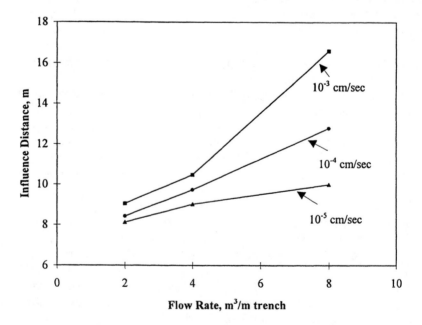

Figure 7.5 Influence distance of horizontal trench.

increases. The simulation was conducted assuming a waste hydraulic conductivity of 10^{-3} cm/sec. At lower hydraulic conductivities, it is expected that greater horizontal spreading will result, however downward movement will be impaired. Consequently horizontal spacing can be increased, however, vertical spacing must be reduced. Influence distance also appears to increase when intermittent leachate introduction is practiced. The influence distances in Fig. 7.5 should be used as a guideline, particularly when placing horizontal trenches near the landfill surface and boundaries.

The saturation isoclines provided by the model (see Chapter 5) suggested that flow rates above 6.0 m³/day/m trench result in the upward propagation of a saturated front and artesian conditions at the landfill surface. The degree of saturation isoclines for the horizontal trench indicate that insufficient spacing of trenches or over-pumping can result in vertical leachate seeps, and possibly artesian conditions. From these results, it is apparent that injection lines should be conservatively distanced from the landfill surface, boundaries, and each other.

At least three meters of waste should be placed on top of a horizontal injection line and a distance of at least 6 m from side slopes to avoid surface uplift, seeps, and artesian conditions. Miller, et al.[101] selected a six-m spacing for trenches based on results of excavation studies at a recirculating cell operated by the DSWA. Intermittent introduction, permitting a fill and drain operation can also prevent saturated conditions from developing. If trenches are placed too close to side slopes or if low permeable daily or intermediated cover is used, French drains should be installed at landfill slopes to control side seeps.

Vertical Recharge Wells

Saturation profiles for the vertical well (see Chapter 5) suggest that the leachate will initially show preferential flow vertically along the well surface. Such flow may contribute to the localized subsidence around the wells at full-scale sites. Higher saturations (greater than 0.6) initially develop along the well surface and slowly begin to propagate laterally and vertically as leachate attempts to percolate downward more quickly than it can be conveyed by the waste matrix. Modeling found that the impact area of a vertical well was a function of the rate of flow, like the horizontal trench. However in this case small increases in flow resulted in large increases in impact area. Vertical well spacing is conventionally 35 to 100 m (118 to 333 ft). Al-Yousfi[78] proposed a simplified estimate of vertical well influence radius based on the relative hydraulic conductivity of the well filling (usually gravel) and the waste, as shown in Equation 7.4.

$$R = \frac{rK_w}{K_r} \tag{7.4}$$

where: R = radius of influence zone, L
           r = radius of recharge well, L
           K_w = permeability of media surrounding well, L/T, and
           K_r = permeability of refuse, L/T.

For example, for a ratio of hydraulic conductivities of 30 to 50 and a well diameter of 120 cm (4 ft), the influence distance would range from 36 to 60 m (120 to 200 ft). A well spacing of 60 m (200 ft) would therefore ensure that adequate wetting would occur. This approach neglects the impact of leachate flow rate on influence distance, however.

As seen in Chapter 5, the vertical well is inefficient at wetting the upper portion of the landfill while the horizontal trench is less effective at wetting the lower portion of the fill, except at high flow rates. A combined system would then deliver the most uniform wetting of the waste with the well impacting the lower portion and the trench impacting the upper portion. Alternative placement of layers of trenches at right angles to each other at each lift may also increase impact area. Use of low permeable material, such as gravel, extending from the vertical wells to serve as wicks are being used at several landfills to increase the impact area.

Design Approach

Thus, the design of a recirculation system should involve more than one recirculation system. Initial wetting with leachate as the waste is placed is also recommended. Once sufficient waste is in place (3 to 6 m [10 to 18 ft]), horizontal trenches and vertical wells can be utilized. Trench spacing can be determined using the following procedure involving Fig. 7.5.

1. Determine desired volume to be recirculated using hydrologic modeling results, i.e., HELP.

2. Determine dimensions of the landfill.

3. Select length of trenches.

4. Determine the flow rate per length of trench.

5. Determine the spacing of the trench.

If flow is below the curve on Fig. 7.5, leachate should be pumped to the trench during portions of the day and rotated from trench to trench to increase flow to each trench. Low flow rates will result in incomplete wetting of the landfill. If the flow is above the curve on Fig. 7.5, the spacing of the trenches should be reduced. Note that the influence distance should be doubled to determine trench spacing requirements.

As an example, consider a 2-ha (five-acre) landfill cell recirculating an average of 15 m³/ha/day. The landfill is 300 m by 67 m and the perforated section of the trenches will be 50 m long. Leachate will be pumped to each trench for four hours each day, resulting in a flow rate of 3.6 m³/hr/m. The influence distance is therefore three m, from Fig. 7.5, and the trenches should be spaced six m apart for a total of 50 trenches.

■ FINAL AND INTERMEDIATE CAPS

The typical landfill today is closed within two to five years, and, because of high construction costs, is often built to depths of well over 100 m (300 ft). Once the landfill reaches design height, a final cap is placed to minimize infiltration of rainwater, minimize dispersal of wastes, accommodate subsidence, and facilitate long-term maintenance. The cap may consist (from top to bottom) of vegetation and supporting soil, a filter and drainage layer, a hydraulic barrier, foundation for the hydraulic barrier, and a gas control layer (Fig. 2.3).

The rapid closure and deep construction of the modern landfill tends to minimize surface exposed to infiltration. Also, many states prohibit the disposal of yard waste in landfills, eliminating an important source of moisture. As a consequence, calculations show that even if emplaced waste captures every drop of precipitation in the wettest climates, moisture content of the waste at closure may still be below optimum levels for biological degradation.[102] In addition, a fraction of the water entering a landfill finds highly permeable pathways to the collection system and is not absorbed by the waste. Once closed, further introduction of moisture is prevented by impermeable caps. As degradation proceeds, moisture content continues to decline with losses to biological uptake, and gas production and waste degradation rates may continue to decrease. These factors contribute to dry-tombing and storage of waste as opposed to rapid stabilization promoted by the bioreactor approach. Data gathered by the Governmental Advisory Associates, Inc[103] showed a strong positive correlation between gas generation at landfills with permeable covers and rainfall. Gas generation at a Spokane, Washington landfill fell dramatically following the installation of an impermeable cap.[104]

Subtitle D of RCRA requires provision of a final cap that will **prevent** the infiltration of precipitation. Postponement of final RCRA closure should be considered in lieu of an intermediate cap (composed of more permeable soil) which provides for limited infiltration of moisture, along with leachate recirculation, to maintain appropriate conditions for biodegradation of waste. The intermediate cap has advantages related to subsidence accommodation as well. Subsidence is discussed further in Chapter 8. Provision of an intermediate cap may require more confidence in the bioreactor landfill prior to acceptance by regulators.

■ GAS COLLECTION

As discussed in Chapter 6, several laboratory and pilot-scale lysimeters have documented increased gas production rates and total yields as a result of moisture addition.[2,38,32] Limited data also suggest that, as in lysimeters, gas production at larger sites is significantly enhanced over traditional landfilling and gas collection practices. The increase in gas production results from accelerated waste stabilization as well as the return of organic material in the leachate to the landfill for conversion to gas (as opposed to washout in conventional landfills). Gas production enhancement can have positive implications for energy production and environmental impact, however, only if gas is managed properly. The facility must be designed to anticipate stimulated gas production, providing efficient gas capture during active phases prior to final capping. Captured gas, in turn, must be utilized in a manner that controls the release of methane and nonmethane organic compounds and provides for beneficial offset of fossil fuel use.

Horizontal collection systems are gaining in popularity as an efficient method of gas extraction for active landfills and, in particular, for bioreactor landfills. The gas extraction trench is constructed in a similar manner to the horizontal injection trench for recirculation: trench excavation, backfilling with gravel, and placement of a perforated pipe. Horizontal spacing of extraction trenches ranges from 30 to 120 m (100 to 400 ft), vertical spacing ranges from 2.5 to 18 m (8 to 60 ft) or one trench for every one or two lifts of waste. Trench design should consider the overburden of continued filling and the effects of compactors moving over the trench.

Gas production has been observed to be particularly enhanced near leachate reintroduction sites. Accordingly, horizontal leachate injection pipes are also being used to extract gas (see Fig. 7.6) at the Alachua County, Florida, Southwest Landfill to maximize gas collection efficiency.[56] Use of leachate introduction devices for gas extraction is being considered at many landfills. Miller and Emge,[100] however, reported problems associated with two-phase countercurrent fluid flow in leachate distribution/gas collection lines. Leachate tended to collect at the bottom of horizontal trenches, blocking the transmission of gas. To avoid this problem, Cell 3 at DSWA's Southern Solid Waste Management Center will integrate leachate distribution and gas extraction by locating respective trenches five feet apart, however gas and leachate transport will be accomplished separately.

A recent study investigated the economics of active landfill management with respect to gas utilization.[105] Three management alternatives were investigated: conventional single-pass leachate operation, bioreactor technology for gas production enhancement, and triggered operation. Triggering employs methods to control the timing of the onset of landfill gas generation such as temperature management, moisture, and nutrient

Figure 7.6   Leachate injection and gas extraction using common horizontal trench.

management.[106] With such control, the majority of gas could be generated following landfill closure and the installation of the gas collection system. While triggering is not yet practical at full-scale it is certainly a desirable outcome of current research efforts.

In the study, bioreactor landfills were found to have advantages of increased waste stabilization and landfill gas production over a shorter period of time. However, if landfill gas is not collected immediately, much of the advantage of energy recovery is lost, and associated safety and environmental risks escalate. In addition, the rapid increase and decline in gas generation associated with bioreactor operation may make it difficult to match gas utilization with gas collection, particularly when using constant capacity alternatives, as well as those requiring expensive cleanup such as electric power or vehicular fuel generation. Thus, much of the gas collected during peak production periods may be wasted. Staging of construction of gas utilization equipment over time may be possible, but only if justified by sustained net revenue over an extended project life. As discussed above, use of horizontal collection systems prior to landfill closure was found to have significant economic benefit over less efficient collection systems when used to extract gas generated by bioreactor landfills. Triggering of gas production was found to have significant economic benefit when compared to conventional operation.

■ CELL CONSTRUCTION

For economic reasons, the recent trend in landfill construction is to build deep cells to provide a life of two to five years. This trend also has certain advantages related to bioreactor design. Cell construction can incorporate latest technological developments rather than committing long-term to a design that may prove to be inefficient. Small, hydraulically separated cells are easier to isolate to minimize stormwater contamination and shed water more efficiently when covered. Baetz and Byer[107] calculated that as much as 30 percent more leachate is generated from horizontal cell construction as compared with extreme vertical construction with minimal face exposure. Once closed, methanogenic conditions within the cell are optimized and gas production and collection is facilitated. It also may be possible to then use the closed cell to treat leachate from new cells as discussed in more detail in Chapter 8. Deep cells improve compaction and anaerobic conditions are more readily established, however, moisture content in small deep cells may be lower than optimum. Therefore, leachate recirculation is essential to efficient waste degradation.

■ CONSTRUCTION COSTS

Most studies show that the cost for implementing the bioreactor technology is minimal when compared to costs for treatment. Many of the components essential to leachate recirculation are common to conventional operation. The leachate pumping station can serve dual purposes of leachate transfer to storage or to reintroduction devices. Leachate storage is always necessary. Consequently the major expense of recirculation is the construction and operation of leachate transmission lines and reintroduction devices.

Construction costs for leachate recirculation at DSWA landfills have ranged from $10,000 to $200,000 (1993 dollars).[89] Onsite leachate treatment costs have been estimated to be $1,000,000 to $6,000,000. Costs for constructing the horizontal injection and spray irrigation leachate recirculation system at the 14-ha (35-acre) Kootenai Count Landfill were $1,035,000 (1993 dollars).[100] When amortized over a 16-year period, this cost is 63 percent less than the cost estimated to transport leachate to a local wastewater treatment facility. Estimated construction costs for recirculation facilities at the Central Facility Landfill in Worcester County, MD, were $26,000 (1989 dollars).

■ SUMMARY

The design of the bioreactor landfill is still an evolving concept, however, there are several features described in this chapter which appear to be essential to the proper utilization of this technology. Many of these features are depicted in Fig. 7.7 and are summarized below:

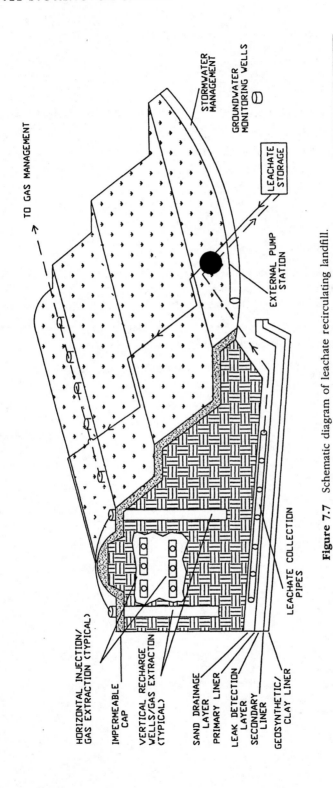

TO GAS MANAGEMENT

STORMWATER MANAGEMENT

GROUNDWATER MONITORING WELLS

LEACHATE STORAGE

EXTERNAL PUMP STATION

HORIZONTAL INJECTION/ GAS EXTRACTION (TYPICAL)

IMPERMEABLE CAP

VERTICAL RECHARGE WELLS/GAS EXTRACTION (TYPICAL)

SAND DRAINAGE LAYER PRIMARY LINER

LEAK DETECTION LAYER SECONDARY LINER

GEOSYNTHETIC/ CLAY LINER

LEACHATE COLLECTION PIPES

Figure 7.7 Schematic diagram of leachate recirculating landfill.

- minimum of a single composite liner comprised of compacted soil and a geomembrane,

- a conservatively designed leachate collection system which will accommodate recirculated flows,

- an appropriately designed drainage and filter system to minimize clogging and head on the liner,

- adequate storage external to the landfill to accommodate storm flow and to provide sufficient flow to continue to recirculate between storm events,

- daily cover which does not affect the passage of moisture through the landfill,

- a small, deep cell active for two to five years,

- a leachate recirculation system which effectively wets the landfill contents, but does not produce side seeps or surface flows, and is compatible with the landfill cap,

- an active gas collection system which captures and controls gas emissions throughout the life of the landfill, and,

- a landfill cover that can maintain integrity as the landfill volume decreases.

8 LANDFILL BIOREACTOR OPERATION

■ INTRODUCTION

To successfully operate the bioreactor landfill, it is necessary to control and monitor biological, chemical, and hydrologic processes occurring within the landfill. As has been previously stated, a bioreactor landfill should be approached as a waste treatment system, thus requiring closer attention to system performance relative to traditional landfills. This chapter provides operational guidance as well as information pertinent to the control of bioreactor landfill operations.

■ WASTE CHARACTERIZATION

Waste Composition

MSW composition affects leachate quality, landfill gas composition and quality, waste degradation rates, and resource recovery potential. Composition is a function of characteristics of the generating population. For example, during the period from 1972 to 1987, the population in the United States increased by 16 percent. During that same period, however, MSW discards increased by 28 percent.[108] This increase can be explained by, among other facts, the 34 percent jump in the number of households (due to divorces, elderly population, and later marriages) which resulted in increases in yard waste and discarded appliances, home furnishings, and clothing. In addition, the work force increased by 38 percent (office employment increased by 72 percent) leading to more paper waste and a change in lifestyle that led to use of convenience foods producing more packaging waste and less food waste.[108] More recently a decline in per-capita waste generation has been noted as well as a reduction in the percentage of waste that is placed in landfills as a result of increased source reduction and recycling.[109]

Variations in waste characteristics must be considered as future land-fills are designed and operated. Composition changes in response to new regulations and population characteristics must be foreseen and accommodated. For example, the Office of Technology Assessment has identified as a high priority the need to reduce MSW generation rates and toxicity.[110] Such a reduction would help to minimize waste management costs, improve the efficiency of the use of natural resources, generate public confidence in governmental MSW policies, and enhance the quality of landfill leachate and gas. As another example, many states are precluding yard wastes from lined landfills, and eliminating a source of moisture, nitrogen, and organisms capable of degrading waste. The impact of these changes on bioreactor operation is difficult to predict.

Preprocessing of wastes does permit some control of the composition of landfilled MSW. Separation of inert and organic waste, bag opening, shredding, and household hazardous waste removal provides for a more uniform waste, improves leachate and gas quality, equalizes subsidence (facilitating post-closure care), and simplifies landfill operations.

Waste Physical Properties

Physical properties of MSW provide some opportunity for reactor control. These properties include in-place density and particle size, which primarily influence moisture routing within the landfill. In-place density can be controlled by compaction in the field or by baling of wastes before landfilling. Field compaction is accomplished by moving heavy equipment over the wastes a number of times. Density increases with the number of passes (three or four is optimal) and decreases with the thickness of the lift. Typical densities vary from 475 to 830 kg/m^3 (800 to 1400 lb/yd^3). Baling can increase density to as high as 890 kg/m^3 (1500 lb/yd^3). Greater compaction (and resulting greater density) has advantages associated with more efficient use of air space, reduced settlement, and reduced cover material requirements. However, hydraulic conductivity is diminished, moisture distribution is impaired, and leachate short circuiting is promoted; therefore leachate may be relatively weak in strength, but waste degradation may be delayed. Successful bioreactor operation requires reduced compaction to promote even leachate distribution. However, increased settlement rates can be expected under these circumstances.

Particle size can be reduced through shredding before waste placement. Unless particle size is reduced to sub-micron diameters, however, shredding cannot overcome mass transfer limitations of degradation, particularly in a dry landfill.[77] Shredding does promote a more uniform waste and reduces fire potential, blowing wastes, and the need for daily cover. Shredding also improves water distribution and promotes more equitable settlement. In addition, more waste is exposed to microbial activity and consequently biodegradation may be enhanced. Dewalle et al.[111] found that shredding significantly increased gas production and concluded that approximately 50 percent of potentially degradable organic material is protected in a landfill without shredding. Bookter and Ham[112] also concluded that shredding increased the rate of decomposition in test cells, as did Otieno.[37] However Tittlebaum[35] found shredding had no effect on degradation in laboratory-scale lysimeters. Reinhardt and Ham[65] reported a 27 percent increase in density for milled refuse at equivalent compaction as compared with nonshredded waste. Leachate from shredded waste was more highly contaminated during early stages and less contaminated during later phases. Reinhardt and Ham concluded that leachate flowed more evenly in milled refuse. Shredding adds significant cost to landfill operations and is more frequently used to facilitate resource recovery and combustion, however, if used judiciously, shredding may provide sufficient advantages in terms of enhanced gas production and minimized long-term liability to justify the cost.

■ OXIDATION REDUCTION CONDITIONS

Oxidation reduction (redox) conditions within the landfill establish waste degradation pathways. High redox potential associated with aerobic conditions provided for accelerated waste stabilization and reportedly improved leachate quality.[52] Aerobic landfilling is more akin to today's composting operations and has been promoted by many investigators since the 1960s.[113]

Aerobic conditions have been encouraged in studies in West Germany using thin-layer (less than 0.4 m) compaction.[52] Thin-layer compaction was found to provide increased and more uniform waste density, improved water distribution, and enhanced penetration of air. The biological process was then accelerated, with easily degraded organic materials removed before the onset of anaerobic conditions, thus limiting the production of organic acids that can retard methanogenesis. Thin-layer compaction was discussed in more detail in Chapter 4.

Studies in the United Kingdom have investigated air injection into completed landfills (using vertical gas wells), in order to promote aerobic degradation.[114] Because of the energetics of aerobic metabolism, temperature temporarily increased locally (by 17°C) and dramatically stimulated subsequent anaerobic activity as evidenced by increased gas production.

American Technologies, Inc. is conducting a pilot-scale experiment with aerobic landfilling at the Baker Place Road landfill — Columbia County, Georgia.[115] Air is injected into leachate collection system clean-out lines and distributed across the landfill through the gravel drainage layer. Three seven-hp single-phase electrical blowers were used in the first phase of the test, operated at 16 sm^3/min (550 scfm) and a pressure of 13 cm water column (5 in). Measurable air flow was detected over 300 m (1,000 feet) away from the blowers. Subsurface vapor monitoring was provided to measure oxygen, methane, and carbon dioxide. Significant oxygen depletion was reported with 60 percent reduction in methane concentrations. Air injection is combined with leachate recirculation in order to maximize waste stabilization rates and minimize closure requirements.

It should be noted that the presence of air in a landfill may create fire potential, has additional operating costs associated with provision of air, and may still produce off-gases that require collection and treatment. Anaerobic degradation, however, leads to the production of methane which can be recovered for energy generation. In addition, anaerobic degradation pathways are available for many compounds that are not amenable to aerobic degradation — for example chlorinated aliphatic hydrocarbons.

Another process that controls oxidation reduction potential within the landfill was investigated at pilot scale in Sweden.[58] This test examined a two-step degradation process, whereby acidogenic conditions are maintained in one portion of the landfill, and methanogenic conditions elsewhere. The high strength, low pH leachate produced within the acid phase area was recirculated to the methane producing area for treatment. Consequently, methane production can be accomplished under controlled conditions in an area suitably designed for maximum methane recovery.

■ MOISTURE CONTENT

Moisture addition has been demonstrated repeatedly to have a stimulating effect on methanogenesis,[31] although some researchers indicate that it is the movement of moisture through the waste as much as it is water addition that is important.[34] Moisture within the landfill serves as a reactant in the hydrolysis reactions, transports nutrients and enzymes, dissolves metabolites, provides pH buffering, dilutes inhibitory compounds, exposes surface area to microbial attack, and controls microbial cell swelling.[77]

Leachate recirculation appears to be the most effective method to increase moisture content in a controlled fashion. The advantages of leachate recirculation include positive control of moisture content, leachate treatment, and liquid storage and evaporation opportunities. Pohland[2] suggested that leachate recirculation could reduce the time required for landfill stabilization from several decades to two to three years. Suflita et al.[90] and Miller et al.[85] both noted the important role of moisture in supporting methanogenic fermentation of solid waste when examining samples removed from operating landfills. Recommended moisture content reported in the literature ranges from a minimum of 25 percent (wet basis) to optimum levels of 40 to 70 percent.[31] During lysimeter studies of the effect of moisture content on waste degradation, Otieno[37] observed that complete saturation was not conducive to methanogenesis. Negative impacts of saturation may be due to poor circulation of leachate and the accumulation of volatile organic acids. In addition, saturation leads to side and surface seeps.

Increased moisture content also permits significant storage of moisture within the fill. Rovers and Farquhar[116] determined refuse moisture retention to be between 10 and 14 percent of dry waste by volume above initial waste moisture. A simple analysis of *in situ* storage can be done given incoming waste tonnage and precipitation data. For example, assume waste receipts average 272 metric tons/day (300 ton/day), annual average infiltration is 0.76 m (30 in), and waste is placed at a specific weight of 593 kg/m^3 (1000 lb/yd^3). If this waste has a moisture holding capacity of 15 percent by weight (or 0.25 m^3/metric ton [60 gal per ton]), a total of 68 m^3/day (18,000 gal/day) of moisture can be absorbed. Consequently, an open area of approximately 3.3 ha (8.0 acres) can be supported by leachate recirculation without *ex situ* storage or treatment. *Ex situ* storage still will be required to deal with peak storm events that occur while the landfill is open, as well as to manage leachate recirculated about closed cells. Storage was treated in more detail in Chapter 7.

■ RECIRCULATION STRATEGIES

The practice of leachate recirculation must balance two processes, the biological processes of waste degradation and the hydrologic capacity of the waste (the rate at which leachate moves through the waste) controlled by the waste permeability. Frequency of leachate recirculation, on a practical basis, has historically been dictated by the inventory of accumulated leachate. Operators more or less have sought a place to put large volumes of water. Such practice can lead to saturation, ponding, and acid-stuck conditions, particularly during early degradation phases. Unfortunately, the level of knowledge of bioreactor operations is still fairly limited, particularly at field-scale, however some guidance can be obtained from related studies of anaerobic digestion.

The biological processes occurring in the landfill are largely anaerobic, as described in Chapter 2. These processes are fairly sensitive to environmental conditions such as pH, temperature, toxic compounds, and the presence of oxygen. These factors are discussed in more detail in later sections of this Chapter. Control of the processes in a landfill is similar to that required for an anaerobic digester, however, since a digester is typically a well-mixed reactor, regulation of the digester is much easier to accomplish. Landfill process control can be achieved to some degree by controlling the rate at which moisture is introduced. It is extremely important to introduce leachate slowly before the onset of methanogenesis, while monitoring gas and leachate quality. High flow rates will deplete buffering capacity and remove methanogens. The presence of methane suggests that methanogenesis is occurring. Leachate should be monitored for pH and, more importantly, VOA and alkalinity concentration. Acid stuck conditions are indicated by a high VOA to alkalinity ratio (greater than 0.25) suggesting low buffering capacity. The pH may still be within an optimum range at this point, however the process may be heading for problems.

Leachate Recirculation Frequency

A procedure employed at several successful bioreactor operations involves the rotation of moisture introduction from one area to another, allowing areas to rest between recirculation episodes. This practice facilitates gas movement and minimizes saturation which is particularly important during the early phases of the degradation process.[41,117] Saturation is indicative of stagnant conditions, which have been shown to be detrimental to the landfill process.

Leachate recirculation should be initiated as soon as possible following waste placement to ensure proper moisture content for biodegradation (once sufficient waste is present to absorb the recirculated liquid, perhaps following placement of the first lift). Doedens and Cord-Landwher[118] observed that leachate recirculation initiated at the commencement of operations at full-scale landfills resulted in a more rapid reduction in leachate organic strength than for landfills where recirculation was delayed for up to four years.

Once gas production is well established, leachate can be recirculated more frequently and at greater flow rates. At this point, the rate of introduction of leachate is controlled by the moisture capacity of the waste which, because we are primarily dealing with vertical flow, is a function of the permeability or hydraulic conductivity of the waste. Hydraulic conductivity was discussed in detail in Chapter 5.

To optimize the rate of flow through the landfill as well as the impact area, the practice of short term high-rate leachate introduction may be best during early phases of operation. In this way, increased moisture conditions can be achieved in areas immediately surrounding the recirculation device, lower back pressure will be encountered, and greater areas of the landfill will be wetted. Calculations can be made based on expected hydraulic conductivity, impact area (from Fig. 7.5), and desired depth of wetting. For example if a 50-m horizontal trench is used with a spacing of 5 m, corresponding with a flow rate of 250 m³/d, the area should be wetted within six days. That trench can be abandoned for several weeks and the next trench utilized. Once methane formation commences, the trench can be used more or less as necessary, while watching for side or surface seeps.

Extent of Leachate Recirculation

To maximize waste stabilization, leachate should be recirculated to all parts of the landfill, if possible. Uniform distribution is extremely difficult to achieve, however, and may best be accomplished through prewetting of waste as it is placed in the fill. Non-uniform distribution may have been the biggest problem with early recirculation attempts, leading to short circuiting, ponding, side-seeps, and interference with gas collection. Ideally, minimal compaction should be practiced for the landfill bioreactor to optimize early leachate recirculation. With time, density will increase and conductivity will decline. However at that point the landfill cell will be closed, consequently minimal leachate will be produced and lower recirculation rates will be acceptable. A large amount of settlement will occur under this scenario and should be anticipated when planning for long-term surface maintenance.

When leachate treatment is the primary objective of wet cell technology, leachate recirculation can be confined to treatment zones located within the landfill where appropriate processes are optimized. Use of *in situ* nitrification, denitrification, anaerobic fermentation, and methanogenesis have been proposed to treat leachate, depending on the phase and age of the waste.[119] As described in a previous section, pilot studies in Sweden have successfully investigated a two-step degradation process within a landfill, whereby acidogenic conditions were maintained in one portion of the landfill, and methanogenic conditions in another part.

■ EFFECTS OF WASTE PLACEMENT RATE

A mathematical model was developed to examine the impact of key operating parameters on leachate quality (specifically COD, although the model could be modified to accept any quality parameter). The model is based on a mass balance about one or more leachate recirculating landfill cells and utilizes data presented in Chapter 6 that describe the rate at which leachate COD increases and decreases. Site specific operating criteria can be used as input to the model to evaluate expected leachate quality. The model was successfully validated using data from a recirculating cell operated by the DSWA.[120]

The model was used to investigate the impact of waste placement rate on leachate COD. Waste placement rate was varied from 20 to 70 metric TPD placed in a landfill 50 m by 50 m by 12 m deep. All other parameters were kept constant. Waste density was assumed to be 600 kg/m^3, void volume 0.4, infiltration rate was 1.5 m^3/day, 95 percent of leachate generated was recirculated, initial leachate COD was 6,000 mg/l, and the cell was 75 percent saturated. Waste was assumed to undergo hydrolysis/acidogenesis for 400 days during which the COD rate of increase was 0.004 day^{-1}. After 400 days, methanogenesis begins and the rate of decrease in leachate COD was 0.001 day^{-1}.

The predicted leachate quality as a function of waste placement rate is shown in Fig. 8.1. At lower waste placement rates, the leachate COD concentration peak is higher and COD declines at a lower rate. Also the lower the waste placement rate, the earlier the COD peak occurs. At higher placement rates, the cell is filled faster and the shift from acidogenesis/hydrolysis (which contributes to high COD) to methanogenesis occurs more uniformly throughout the cell. Thus the rapidly filled cell behaves more like laboratory-scale reactors that are filled almost instantaneously and tend to have sharper COD peaks. The slowly filled cells tend to prolong the high COD, acidic phase.

■ USE OF OLD CELLS

The present design of the MSW landfill generally calls for a series of hydraulically separated cells that are opened and closed sequentially.

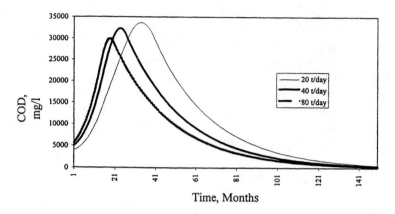

Figure 8.1   Effect of waste placement rate on leachate quality.

Active landfills in most parts of the US generate relatively large volumes of leachate that tend to become increasingly contaminated as waste is placed. Leachate recirculation at this time is important to ensure that the waste reaches moisture levels near or above field capacity. With the provision of appropriate operational controls which include leachate recirculation, the landfill, once closed, can function as a bioreactor, providing *in situ* treatment of organic fractions of the waste as well as recirculated leachate. Within a short period of time following closure, the quality of the leachate will improve, as shown in Fig. 8.2.

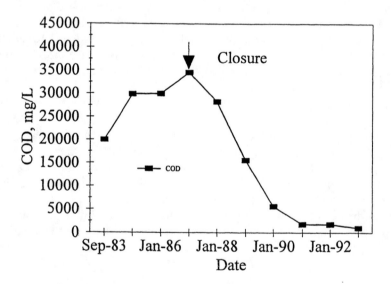

Figure 8.2   Chronological plot of COD data before and after closure of leachate recirculating cell.

As the next cell is opened, leachate volume and strength will increase once again. However, leachate produced from this cell can be recirculated to both the closed cell and the active cell, providing *in situ* treatment of the leachate and moisture control for the active cell as depicted in Figs. 8.3–8.5. Although leachate organic strength will rise and fall with each reactor opening and closing, the magnitude of each subsequent cycle should be dampened as a result of the *in situ* treatment provided by the closed cells. This phenomenon has been observed at the full-scale Lycoming County Landfill that has practiced leachate recirculation since the mid 1970s[49] (see COD data presented in Fig. 8.6) and investigated using laboratory-scale lysimeters by Doedens and Cord-Landwher.[36] In some situations, this practice may be difficult because RCRA Subtitle D precludes the introduction of non-indigent leachate at present.

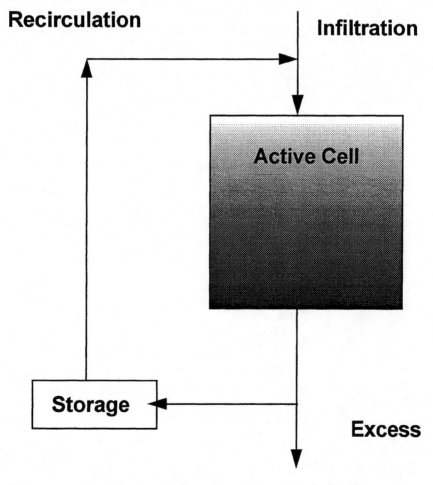

Figure 8.3 Leachate recirculation scenario — single cell.

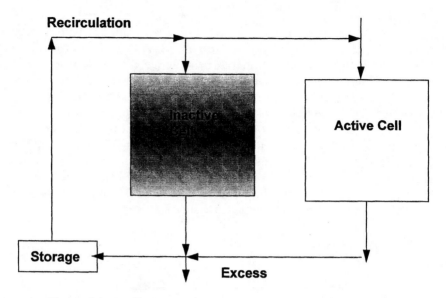

Figure 8.4 Leachate recirculation scenario — two cell sequencing.

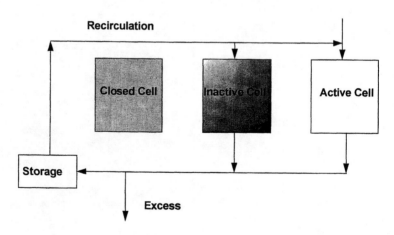

Figure 8.5 Leachate recirculation scenario — closed cell.

The mathematical model described in the above section was also utilized to evaluate the effect of sequential cell recirculation. Two cells were modeled with input parameters as described above. All of the leachate generated by the active cell was recirculated to the inactive cell. The COD output from both a single cell and the two cell configuration is plotted in Fig. 8.7. Leachate recirculation over the inactive cell makes a drastic difference in leachate quality due to the attenuation of leachate organic strength by the fully established methanogenesis phase of the

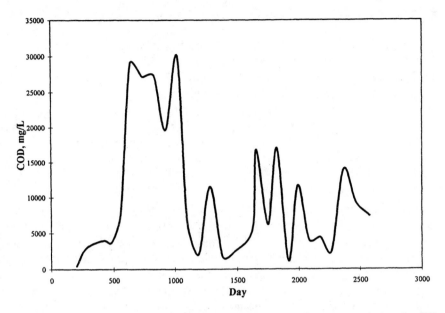

Figure 8.6 Chronological plot of COD data for a leachate recirculating landfill experiencing sequential opening and closure of cells.

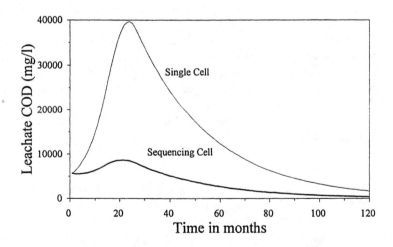

Figure 8.7 Comparison of leachate quality for single and sequencing cells.

inactive cell. Two-stage leachate recirculation can also be achieved through composting or aerobic pre-stabilization of bottom layers of waste in order to treat leachate generated in the top layers of the landfill as it passes through the bottom layers. Clearly, use of this leachate management option should be further evaluated at field scale.

■ BIOREACTOR AUGMENTATION

The consortium of microorganisms involved in the stabilization of waste has specific environmental requirements for pH, temperature, and micro- and macro-nutrients, among others. With the multitude of factors influencing landfill processes, it has been difficult to evaluate, full scale, the impact of any one variable. Laboratory studies have investigated many of these environmental parameters and the results, described in following sections, can be helpful in determining which elements might be important to control. Field studies are in progress to obtain more insight into full-scale landfill effects.

Temperature Control

Because waste degradation involves biochemical reactions, the rate of degradation tends to increase with temperature. The temperature within a landfill is determined by a balance between heat production during the biological degradation of organic waste fractions and the loss of heat to the surrounding soil and atmosphere.[117,121] The microbial processes are capable of significant heat generation, particularly at higher moisture conditions. Optimum temperature has been reported to be 40°C by Gurijala et al.[122] and 34 to 38°C by Mata Alvarez and Martina-Verdure[30] with significant inhibition observed at temperatures above 50 to 55°C.[44] Temperature control at full-scale landfills may be difficult to economically achieve. As previously discussed, introduction of air and the consequential onset of aerobic activity serves to rapidly increase temperature and has been found to stimulate methane production. Another potential method of temperature control under investigation is the heating of recirculated leachate such as used in Sweden's experimental "Energy Loaf".[48] At this point in time, the use of temperature to control the landfill bioreactor is not practical and its benefits not thoroughly demonstrated, therefore it is not recommended.

Nutrients

Nutrient requirements are generally met by the waste at least during early degradation phases,[31] although phosphorous may be limiting during later stages. During methane potential testing of samples retrieved from a DSWA landfill, Pohland[123] found that the addition of nitrogen and phosphorous to mature samples stimulated methane production. Tittlebaum[35] added nutrients to laboratory-scale lysimeters and reported no effect on degradation rates. Mata-Alvarez and Martinez-Verdure[30] indicated that laboratory-scale MSW digesters achieved 90 percent degradation of biodegradable material within 27 days without nutrient addition, provided other environmental factors were within optimal ranges. Nutrient addition adds to the complexity of operation and does not appear to offer sufficient advantages during active landfill phases to warrant its use.

Buffering

Optimum pH for methanogens is approximately 6.8 to 7.4. Buffering of leachate in order to maintain pH in that range has been found to improve gas production in laboratory and pilot studies.[2,35,42,114,124] Gurijala et al.[122] examined samples removed for the Fresh Kills Landfill in New York City for methane potential and found no methane production outside of this range. Buffering as a control option may be best used in response to changes in leachate characteristics (i.e., a drop in pH or increase in VOA concentration) in conjunction with leachate recirculation.[52] Particular attention to pH and buffering needs should be given during early stages of leachate recirculation when excessive moisture can lead to an accumulation of acids, low pH, and inhibition of methanogens. Careful operation of the landfill bioreactor initially through slow introduction of leachate should minimize the need for buffering. Provision for interim lime or sodium hydroxide addition and mixing at leachate storage should be considered.

Inoculation

Seeding or inoculating the landfill has been investigated, usually through the addition of wastewater treatment facility biosolids. In laboratory and pilot studies, biosolids addition has had mixed impact on degradation.[31] ten Brummeler et al.[125] however, found that composted sludge had a positive effect on the start up of MSW digestion, preventing souring experienced in reactors with leachate recirculation alone. Studies by Leuschner,[124] Pohland,[123] and NYSERDA[45] all found positive effects on methane generation onset and rates following the addition of biosolids. Stegmann and Spendlin[52] indicated that biosolids addition does not enhance gas production. Leckie et al.[41] found that the addition of septic tank pumpings stimulated acid fermentation and suppressed methane generation. Any effect measured may be due to buffering or moisture addition more than seeding. In most cases, however, operators prefer to exclude biosolids because biosolids are difficult to work with at the waste face.

Composted waste offers an alternative source of seed. Studies in West Germany found that old refuse was superior to sludge in stimulating methanogenesis.[52] The addition of a lift of new waste on the top of 1.5 m of five-year-old waste also demonstrated the ability of stable waste to treat new leachate.

■ DAILY AND INTERMEDIATE COVERS

As moisture moves through the landfill, waste heterogeneity in permeability will be frequently encountered, leading to horizontal movement and the potential for leachate ponding or side seeps. The introduction of daily or intermediate cover of low permeability can be particularly troublesome when attempting to introduce large volumes of leachate to the site. Outbreaks and ponding of leachate were reported at the Seamer Carr Landfill investigation.[99] Accumulations within the site to depths of 1.2 m (48 in) or more were reported due to the use of low permeability soil cover. Natale and Anderson[49] also reported saturated conditions and ponding at the Lycoming County site during periods when high volumes of leachate were recirculated in areas using clay and silty soils for daily cover. Lechner et al.[46] observed ponding within the Breitnau test cells and attributed it to calcium deposition. Exhuming portions of a recirculating cell operated by the DSWA revealed perched water within the landfill as a result of layers of cover that prevented moisture movement.[101] Waste in dry areas beneath perched water showed almost no degradation.

In order to minimize ponding and horizontal movement, use of high permeability soils and/or alternative daily cover should be considered. Alternative daily cover materials include mulched or composted yard waste, foam, carpet, clay/cellulose additives, and geotextiles. Geotextiles and carpet should be removed before the next addition of waste, while the clay/cellulose additives crumble and foam quickly dissipates when waste is placed on top of it. In either case, the flow of moisture is not impeded by these materials. The use of geotextiles offers an advantage over other cover materials in controlling the rate of precipitation infiltration into the landfill. Analysis shows that alternative daily cover can be cost-competitive with natural soils.

■ SETTLEMENT

Another consequence of bioreactor operation is enhanced settlement rates. Settlement is caused by the following factors:[7]

- reduction in voids space and compression of loose materials due to overburden weight,
- volume changes due to biological and chemical reactions,
- dissolution of waste matter by leachate,
- movement of smaller particles into larger voids, and
- settlement of underlying soils.

Studies investigating the impact of leachate recirculation on settling have shown that wet cell technology enhances the rate and extent of subsidence. At the Sonoma County, CA, pilot-scale landfills, the leachate recirculated cell settled by as much as 20 percent of its waste depth, while dry cells settled less than 8 percent.[41] Wet cells at the Mountain View Landfill, CA, settled approximately 13 to 15 percent, while control dry cells settled only 8 to 12 percent over a four- year period.[32] Wetting of waste as it is placed has been practiced for many years as a method of increasing compaction efficiency. Rapid and predictable settlement can provide an opportunity to utilize valuable air space before closure of the cell. Enhanced degradation rates can provide a means to meet mandated waste volume reduction in some parts of the world. Landfill reclamation and final site use are also facilitated by timely volume reduction provided by moisture control. The difficulty and expense of long-term final cover maintenance can be reduced as well. One potential problem with enhanced settling is the impact on leachate reintroduction and gas collection pipes and valves. Appropriate use of flexible leachate feed connections to trenches and wells is recommended to prevent pipe breaks.

Settlement also may negatively impact the integrity of internal leachate recirculation devices. Vertical wells may experience displacement due to waste shifting. Trenches will experience variable settling rates and consequently pipe breakage and low spots may occur. In these cases leachate distribution may be uneven. Some designers have recommended telescoping vertical piping to accommodate settling, however this approach does not accommodate horizontal shifting. It is anticipated that use of sufficiently permeable trench backfill will provide continued service even if pipes break, clog, or settle. Therefore, great care should be taken in selecting and placing backfill material. It may be necessary to retrofit the landfill with surface infiltrators and vertical wells after the majority of settlement has occurred, perhaps in conjunction with final closure construction. Again, more information is needed to evaluate the long-term survivability of recirculation devices.

Settlement can be controlled by creating a stable foundation between the waste layer and the cover.[126] The foundation layer may contain 0.6 m (2 ft) or more of compacted coarse-grain material. This layer serves several functions in addition to minimizing settlement, including a subbase to the cap and a gas control layer. Increasing the thickness of other layers within the cover (barrier, erosion control, and vegetation layers) can also reduce settlement problems.

■ MONITORING

Landfill monitoring is important from a regulatory, operating, and design perspective. Use of the landfill as a bioreactor necessitates additional monitoring efforts because it still is considered innovative and because monitoring facilitates control of the process. Augenstein and Yazdani[127] recommends monitoring of gas recovery and composition, waste characteristics, leachate flow and composition, liner integrity, and waste settlement. Table 8.1 provides recommendation for bioreactor process performance monitoring. McBean et al.[7] provides a more general monitoring program for landfill operation.

TABLE 8.1

Bioreactor Process Performance Monitoring

Parameter	Frequency	Ex Situ Leachate Treatment	Environmental Impact	In Situ Leachate Treatment	Gas Control
Leachate					
Organic strength	Monthly	×	×	×	
Volatile organic compounds (VOCs)	Quarterly		×		×
Synthetic organic compounds (SOC)	Quarterly		×		
Metals	Quarterly	×	×	×	
Nutrients (phosphorous, ammonia)	Quarterly	×	×	×	
Nonmetal inorganics (TDS, sulfate, chloride, potassium)	Quarterly	×	×		
Volatile organic acids	Monthly	×	×	×	
Flow rate	Continuous	×	×	×	
Gas					
Methane, carbon dioxide	Weekly		×		×
VOCs	Quarterly		×		×
Hydrogen Sulfide	Quarterly		×		×
Flow Rate	Continuously				×
Groundwater: Quality	Quarterly		×		
Waste					
Cellulose/Lignin Ratio	Twice annually			×	
Ash Content	Twice annually			×	
Biological Methane Potential	Twice annually			×	
Appearance	Twice annually			×	

Purpose of Monitoring

Measurement of gas flow rate can be particularly troublesome because of high moisture and contamination. A number of instruments are available including the pitot tube, orifice, venturi, vortex shedding, and thermal dispersion meters. An excellent review of available gas meters is provided by Campbell.[47]

Waste characteristics are also difficult to assess *in situ*. Soil type moisture sensors (gypsum blocks) have been used in several installations recently to provide continuous moisture monitoring, however it is still too early to conclude on their effectiveness. It is expected that gypsum blocks will have a short life due to leachate attack and may be most useful in indicating moisture arrival rather that exact moisture content. Pressure transducers placed on the liner have also been used to monitor hydrostatic head. Other important waste and leachate analyses include cellulose, lignin, and ash content, biological methane potential, nutrient content, and alkalinity. These parameters can be useful in determining the relative stability of the waste. Sample retrieval is discussed by Miller et al.[85] using 15-cm (six-inch) augers and Suflita et al.[90] using a bucket auger.

Leachate characteristics can provide important insight relative to waste degradation phase. Analysis therefore should reflect the expected reactions occurring within the landfill. BOD, COD, pH, VOA, TOC, nutrients, and alkalinity are important early in the life of the cell. COD, metals, ammonia, and conductivity are important at maturity to develop plans for inactivating the landfill and treating remaining leachate.

■ WHEN IS THE WASTE STABLE?

Once waste stabilization has been satisfactorily achieved, the cell should be deactivated by discontinuing leachate recirculation and removing all liquid. The liquid undoubtedly will require some form of physical/chemical treatment to remove remaining recalcitrant organic compounds and inorganic contaminants. The point at which degradation is sufficiently complete to consider the landfill "stable" is not clearly defined. The most appropriate indicators appear to be leachate quality, gas quantity, and waste composition. Consideration of pilot-scale testing of leachate recirculation at the Georgia Institute of Technology[38] offers some insight into this question. In the presence of leachate recirculation, organic strength of leachate, as measured by COD, reached a minimum level approximately 1000 days after startup. However, at this time, only 68 percent of the total gas production had occurred. When over 95 percent of the gas had been produced, gas production rates had fallen well below 10 percent of peak rates. Thus, it appears from this study that sufficient waste stabilization occurs when gas production reaches relatively low rates (less than 5 percent of peak value) **and** leachate strength remains low (COD below 1000 mg/l, BOD below 100 mg/l). Other indicators are low leachate

BOD/COD ratio (less than 0.1), waste cellulose/lignin ratio (less than 0.2),[112] low waste biological methane potential (less than 0.045 m³/kg volatile solids added),[128] and dark and sludge-like appearance of the waste. While examining exhumed waste from a DSWA landfill, Pohland[123] concluded that low volatile solids content of the waste was a misleading parameter with respect to waste stability.

9 MATERIALS RECOVERY AND REUSE FROM BIOREACTOR LANDFILLS

■ INTRODUCTION

The previous chapters in this book have focused on the operation of the modern sanitary landfill as a bioreactor to promote rapid stabilization of the landfilled waste. The primary end product from the treatment of waste in a bioreactor landfill, in addition to LFG, is the stabilized residual material. This material consists of a mixture of cover soil, the remnants of the biodegradable organic waste, and non-biodegradable waste components such as glass, plastic, and metal. The composition of the stabilized landfill material is a function of the original waste composition and the amount of cover soil placed in the landfill unit. The degraded organic waste possesses similar physical characteristics and properties as compost from other biological solid waste treatment operations such as aerobic composting and in-vessel anaerobic digestion. The degraded organic fraction in combination with cover soil — which may be a sizable percentage of the total mass of landfilled material — represents a potentially recoverable material.

The reclamation of stabilized landfill residual from a treated landfill is a logical final step in bioreactor landfill operation.[129] A landfill is operated as an anaerobic treatment vessel and the final product is reclaimed after treatment. Recovery not only includes the stabilized residue and other recyclable materials (primarily ferrous metal), but valuable landfill space is reclaimed for future waste disposal and treatment operations as well. Future waste management strategies could consist of a series of bioreactor landfill units operated to rapidly stabilize the waste, collect LFG as a beneficial byproduct, and ultimately recover and process remaining materials, thus extending the life of a landfill site many years beyond the traditional time period.

This chapter explores the concept of landfill reclamation and the recovery of materials and landfill volume as the final step in bioreactor landfill operation. This post-treatment option is by no means a necessary part of landfill operation as a bioreactor. Most current applications of the technologies presented in this book involve stabilization of a landfill that remains in place after closure. The potential for reuse and recovery, however, makes the concept worth additional discussion. It is important to note that the system outlined here is *not* one which is in active practice

155

as an integrated waste management system in the United States. As previously discussed, many facilities practice leachate recirculation to stabilize the landfilled waste. In addition, many landfills have been reclaimed or *mined*, and this technology has gained considerable experience in recent years. This chapter therefore outlines the concept of integrating existing technologies to both treat and recover solid waste in bioreactor landfills.

■ LANDFILL TREATMENT AND RECLAMATION STRATEGIES

Benefits of Bioreactor Landfill Treatment and Reclamation

The advantages of bioreactor landfill operation relative to traditional landfill operation have been previously discussed. Rapid waste stabilization reduces long-term costs and risks resulting from landfill operation. Questions must be addressed, however, regarding the benefits accrued by the reclamation of the landfilled material. One notable benefit is the recovery of landfill volume. The expense and effort involved in the siting, design, and construction of modern sanitary landfills makes disposal capacity a valuable commodity. A landfill that is excavated to reduce the mass and volume of material which must be permanently land disposed presents an opportunity for reuse of the landfill capacity. Reclamation activity also permits an inspection of the integrity of the landfill liner and leachate collection system if needed. Since much of the cost associated with modern landfill operation derives from closure and post-closure monitoring and maintenance, a reduction in the volume of permanently landfilled material may produce substantial savings.

When the treatment of solid waste in order to avoid permanent land disposal is a waste management goal, other options are available, including incineration, aerobic composting, and in-vessel anaerobic treatment. In some situations, the use of bioreactor landfill treatment systems may offer a number of advantages over these technologies. One advantage is derived from the fact that these other technologies are, in most cases, more costly than the modern sanitary landfill, even one modified for bioreactor treatment. While incineration of waste results in the greatest destruction of waste material, use of waste combustion is encountering ever increasing difficulty associated with permitting and public acceptance.

Treatment of waste in a landfill bioreactor ultimately produces the same end-product as other forms of biological treatment — a stabilized compost-like material. Aerobic composting and in-vessel anaerobic digesters offer more opportunity for control over the process and much more rapid treatment. This control and rapid treatment, however, requires greater expense in capital equipment and operation. Rapid treatment times also result in a constant supply of residual material that requires a stable reuse market. One of the major impediments to the success of full-scale

MSW composting systems has been the availability of these markets. Consequently, large stockpiles of this residual material can be found at some facilities. Unfortunately, this material often is ultimately landfilled when a market cannot be found. In the case of the bioreactor landfill, if markets do not exist for the material, the landfill provides secure, long-term storage within a liner and odor control system.

The issue of potential reuse options for recovered material from reclaimed landfills will be discussed in Section 9.5, however, it should be noted that the quality of treated waste from any biological treatment system is a function of the characteristics of waste entering the facility. An aerobic composting facility and a bioreactor landfill facility which receive the same waste stream, will likely produce the same quality of stabilized organic matter. The primary difference will be that the residual from the bioreactor landfill will be mixed with cover soil. While it is perhaps true that landfills may be more likely to receive small amounts of hazardous materials, this trend is in decline as a result of more aggressive household hazardous waste diversion programs as well as waste inspection.

Conceptual Operation Approach

The modern landfill unit is the foundation of the reclamation system. Leachate recirculation may be practiced to start the rapid stabilization process while waste is being added to the landfill. It is unlikely that treatment will be complete when the landfill reaches capacity, therefore, an additional treatment period is required, necessitating that multiple bioreactor landfill units be used. This section illustrates the use of this technology conceptually. In this scenario three units are considered. Initially, one bioreactor unit receives waste and cover soil (Fig. 9.1). While the bioreactor is filled, leachate recirculation and gas extraction are underway. When Unit 1 reaches capacity, a second unit begins operation (Fig. 9.2). While Unit 2 accepts waste and cover soil, Unit 1 is actively treated, and gas is withdrawn and processed. The system is sized so that when Unit 2 reaches capacity, Unit 1 waste has been stabilized and is ready for reclamation. At this point, Unit 3 begins operation, Unit 2 waste is treated, and the reclamation of Unit 1 begins (Fig. 9.3).

The process of reclamation generates both a fine fraction consisting of cover soil and degraded waste, and an oversized fraction. The oversized fraction may be processed for other recoverable materials if desired. If the fine fraction meets applicable regulatory requirements for beneficial reuse outside the landfill environment, this material would be diverted from permanent land disposal. As an alternative reuse option, this material could be used as cover soil in the current bioreactor unit. This practice would alleviate the need to excavate virgin soil, further extending the life of the site. The oversized material that could not be recovered would be re-disposed. This fraction could either be permanently stored in a

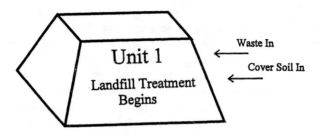

Figure 9.1   Landfill bioreactor with reclamation: phase I.

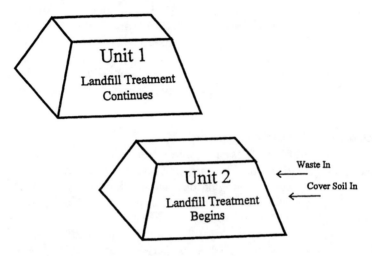

Figure 9.2   Landfill bioreactor with reclamation: phase II.

specified area of one of the treatment units, or in a separate residual disposal unit designated for stabilized oversized material. The stabilized nature of this fraction may offer the opportunity to employ less stringent and expensive liner designs similar to those required for disposal of inert wastes like construction and demolition debris.

■ MASS BALANCE DESIGN FOR LANDFILL RECLAMATION

Once a conceptual model for a bioreactor with reclamation has been developed, it is necessary to perform a mass balance analysis to size the bioreactor landfill units, to determine the amount of reclaimed material, and to estimate the composition of the reclaimed material.

Bioreactor Landfill Unit Sizing
The required size of a bioreactor landfill unit may be determined in the same fashion as a traditional landfill with the requirement that the capacity

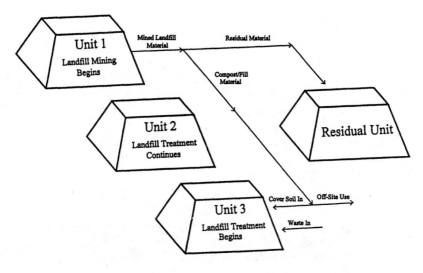

Figure 9.3   Landfill bioreactor with reclamation: phase III.

(in years) must be based on the time necessary for waste treatment to occur sufficiently to permit reclamation. For the three-bioreactor unit system previously outlined, waste from one unit at a time will undergo treatment while the other two units will either be actively accepting waste or being reclaimed. Some treatment will occur while the waste is placed and during the reclamation process, but it is assumed here that the time to treat the waste must be provided during the active treatment phase. The size of the bioreactor landfill, $V_{BIOREACTOR}$, may be determined using Equation 9.1.

$$V_{BIOREACTOR} = \frac{M_{WASTE}}{\rho_{WASTE}}\left(1 + CC\right)t \tag{9.1}$$

where: M_{WASTE} = Incoming waste deposition rate (mass/time)

M_{WASTE} = Density of the waste (mass/volume)

CC = Cover soil content defined as the ratio of the volume of cover soil to the volume of waste, and

t = Time it takes to fill the bioreactor. This should correspond with the treatment time required for the unit that undergoes active treatment.

The Amount and Composition of Reclaimed Material

As previously stated, the stabilized landfill material will include a residual fraction composed of cover soil and degraded organic material, and a large fraction composed of non-degraded materials. The relative amount of each material can be estimated from the incoming waste composition and the amount of cover soil used. The composition of the incoming waste can be based on local waste composition studies or state and national averages. The waste components are categorized as either biodegradable or nonbiodegradable. Nonbiodegradable components remain at the end of landfill stabilization. It is important to note that the biodegradable components do not degrade completely, and a residual stabilized material remains. The relative fraction of material that is degraded and converted to LFG is defined here as BF, and is a function of the type of material. Some types of paper, for example, have a higher lignin content that results in less biodegradation, and more mass remaining after decomposition. Table 9.1 presents the BF of a number of biodegradable landfill materials.

TABLE 9.1

Biodegradable Fraction of Selected Waste Components[4]	
Waste Component	**BF**
Newsprint	0.21
Office Paper	0.79
Cardboard	0.44
Yard Wastes	0.36 to 0.65

The relative amounts of materials before and after stabilization are presented for a typical waste stream in Fig. 9.4 on a mass basis. The percentage of soil by mass is much greater than the percentage by volume because the density of cover soil is typically greater than that of the waste. Typical compacted densities of waste range from 600 to 700 kg/m^3 (1000 to 1200 lb/yd^3), while the density of soil is approximately 1600 kg/m^3 (2700 lb/yd^3).

The rate of material to be reclaimed from the fill will be subject to the strategy and goals of the bioreactor landfill operator. If a goal is to match the amount of waste entering the site (assuming a constant rate of waste generation), the rate of landfill reclamation can be expressed using Equation 9.2:

$$R = M\left(\frac{\rho_{SOIL}}{\rho_{WASTE}}CC + f_{NDF} + f_{DF}(1 - BF)\right) \qquad (9.2)$$

where: R = Reclamation rate of mined material (mass/time)
      f_{NDF} = Fraction of dry mass which is nondegradable
      f_{DF} = Fraction of the mass which is degradable
      BF = Percentage of the degradable fraction that is converted to gas during stabilization

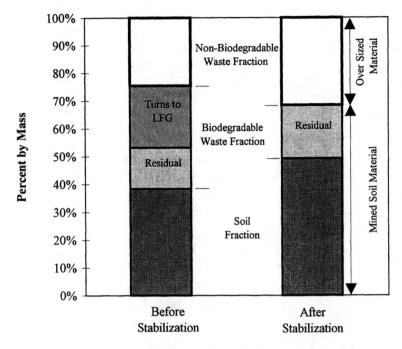

Figure 9.4 Mass balance of landfill material: before and after stabilization.

METHODS OF LANDFILL RECLAMATION

Methods for landfill reclamation have been developed from existing procedures for soil excavation and waste processing. The process of landfill reclamation first involves excavation of landfilled material, which is usually accomplished using a bulldozer or backhoe. The extension of reclamation technology to a bioreactor treatment system requires some estimate of the degree of treatment to determine which areas of the landfill are ready for reclamation. The degree of treatment, as discussed in Chapter 8, may be determined indirectly from measurement of such parameters as LFG production, landfill settlement, and leachate quality, or directly by collecting samples of the waste. Direct sampling of the landfill materials may be conducted by exploratory borings or excavation test pits. Once excavated, the landfill material is then transported to a staging area where it undergoes processing. Recent landfill reclamation activities have taken place on top of or adjacent to the landfilled waste itself rather than at a central processing facility. The equipment is mobile and may be transported around the site as needed.

Following excavation, the landfilled material is processed. The type and degree of processing is dictated by the goals of the reclamation operation. A typical flow diagram for an excavation and processing methodology is presented in Fig. 9.5. Waste is first separated using a large-opening screen such as a grizzly, bar, or finger screen to remove oversized objects. The passing material is then loaded onto a rotating trommel screen. The trommel rotates at a slight inclination, and the fine materials that pass through the screen openings are carried by conveyor belt to a container, truck, or stockpile. Typical screen size opening range from 0.6 to 2.5 cm (0.25 to 1.0 in). A number of manufacturers market screening equipment for landfill reclamation activities. The oversized fraction is carried by a separate conveyor to its own storage area. Additional process steps can be included in order to meet specific goals of the project. Excavation methods for three of the first landfill reclamation projects are discussed in the next section.

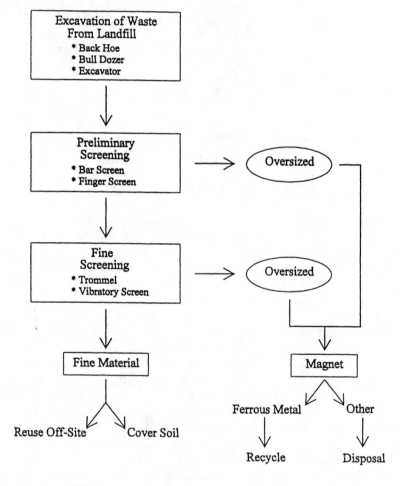

Figure 9.5 Typical landfill reclamation process.

■ PREVIOUS EXPERIENCES WITH LANDFILL RECLAMATION

Landfill reclamation, also referred to as landfill mining, is a technology that has gained considerable practice during recent years.[129] Projects have been conducted in numerous areas in the United States including Florida, New York, Pennsylvania, Connecticut, and Massachusetts, as well as in Germany. The rationale for conducting landfill reclamation projects usually includes the need to remediate old landfill sites, the desire to produce additional fuel for a waste-to-energy facility, and the need to produce cover soil for current landfill operations. Three of the first landfill reclamation programs used in the United States are discussed here and include a description of the operations at facilities in Collier County, Florida; Edinburg, NY; and Lancaster County, Pennsylvania. Other projects have also been implemented at sites in York County, Pennsylvania, Thompson, CT; Kingston, MA; Hague, NY; Horicon, NY; and Colonie, NY.[129-130]

Collier County, Florida

The first landfill reclamation project in Florida was conducted at the Naples Landfill in Collier County, Florida. Landfill mining began on a 11-ha (26-acre) landfill in 1986. Reclaimed waste has been processed at rates ranging from 36,000 to as high as 73,000 kg/hr (40 to 80 tons/hour) at various times during the project. Through October 1996, one third of the original 10 ha (26 acres) was reclaimed. The MSW mined during the operation ranged in age from 10 to 15 years and came from an unlined landfill. The objectives for mining at this site were to recover and reuse cover soil for current landfill operation, reclaim disposal capacity for future landfill activities, reduce the potential for groundwater contamination, recover recyclable materials, and provide supplemental fuel for a proposed waste-to-energy facility.[131]

Between 1986 and 1992, Collier County mined over 63,000 tonnes (70,000 tons) of reclaimed solid waste and cover material. The recovered landfilled material was composed of 75 percent by a fine fraction (soil and decomposed waste) and 25 percent by oversized materials. Excavation was conducted using a front-end loader, a dozer, and an excavator that fed waste to a trailer-mounted coarse screen. Initially, a 7.5-cm (3-in) screen was used to separate the mined material. The material above the screen went to a conveyor belt where the ferrous metals were separated using a magnet. The initial separation system did not meet project needs because the screened passing fraction contained particle sizes that were too large. Ferrous metal separation was also found to be inefficient and the material was excessively contaminated.

A number of research projects were conducted at the site in an effort to increase separation efficiency.[132-133] A trommel screen with 0.6-cm (0.25-in) openings was added to the system. This resulted in recovery of 94 percent of the fine materials. These fines were significantly less contaminated than previously encountered with the 7.5-cm (3-in) screen-ing system. This system was judged to be ineffective in collecting recy-clables, however.[133] Additional waste separation technologies were employed to enhance the collection of plastic, ferrous metals, and alu-minum. After the trommel screen separated the fines, an air knife was used to separate the plastics from the waste stream. A magnet separator collected the ferrous metals followed by an eddy current separator to remove aluminum. This system efficiently separated aluminum from fer-rous metal, but both products contained significant amounts of impurities. A significant amount of fine material was encountered with the ferrous metal fraction and a large amount of wood and plastic was observed in the aluminum. None of these materials met typical recycling criteria for purity. The ferrous metals and aluminum were sold offsite to recycling centers. The remainder of the oversized waste was re-landfilled.

The principal product from the Collier County landfill mining project was the recovered fine fraction of cover soil and decomposed waste (75 percent of the mined material). The recovered soil was used as cover material for active landfill operations at the site. The fine material was compared to existing compost application standards in Florida to examine potential reuse options. The specific characteristics which were investi-gated were the percentage of foreign matter, particle size, and the concentration of heavy metals. The mined material was found to meet all quality standards for heavy metal concentrations for the most stringent Florida compost regulations existing at the time, but the material contained a larger amount of foreign material (between 4 and 10 percent) than typically encountered with compost.

The Collier County project was successful in demonstrating, for the first time, that landfill reclamation could be performed on a relatively large scale. Materials such as ferrous metal and aluminum could be recovered in an efficient manner, but the product quality was lower than typically encountered due to contamination with soil and other fine material. In addition, odor production during the mining process was a serious problem. Synthetic liners were later employed to cover waste and control odor.

Edinburg, New York

The first mined landfill in New York was the Edinburg Landfill, a 2-ha (5-acre) unlined MSW disposal facility.[134] The landfill operated between 1969 and 1991. Prior to 1987, the waste was landfilled using a trench-and-fill technique, creating discrete disposal areas in the landfill and an inefficient use of area. The State of New York chose the Town of Edinburg as the host site for a 0.4 ha (1-acre) landfill reclamation demonstration project. The goal of this project was to develop an economically feasible and acceptable alternative to traditional landfill closure and to reduce the footprint of the landfill. The landfill was mined from fall 1990 to the end of spring 1991. Approximately 0.4 ha (1 acre) of the landfill was reclaimed in 24 excavation days, removing a total of 11,000 m³ (15,000 yd³) of material.

The landfill was excavated using a track excavator with a 1.8-m³ (2.4-yd³) bucket. Front-end loaders were used for minor excavation, material transport, and grading. The landfill was mined using a conventional surface mining technique. The excavated waste was between 11 and 21 years old, and the maximum depth excavated was 6 m (20 ft) with an average excavation depth of 3.4 m (11 ft). Waste depth was considered one of the most critical factors affecting the economics of landfill mining.

Two screening techniques were tested at the landfill. The first technique used a two-stage screening system with a 7.5-cm (3-in) finger screen on top and a 1.3-cm (0.5-in) finger screen on the bottom. The second screening method used a trommel screen with a 23-cm (9-in) scalper for removing larger materials. With the trommel, waste separation occurred on the 3-m (10-ft) long, 1.8-m (6-ft) diameter revolving drum with 2.5-cm (1-in) screen openings. The study concluded that the trommel screen was more efficient than the finger screen system. The trommel oversize fraction contained less fine material and could be fed to a dump truck by conveyor belts. However, the fine fraction from the finger screens possessed less foreign matter, especially glass, than the trommel screenings. Both screening technologies processed reclaimed material at rate of 880 to 960 m³/d (1,150 to 1,250 yd³/d).

The fine material fraction was reused as cover for the existing landfill operation and was stockpiled for later use to construct embankments at the existing landfill. The concentration of heavy metals in the soil met all applicable requirements for compost at the time. The oversized materials, for the most part, were re-landfilled in the active portion of the landfill. White goods and other ferrous metal items were recycled.

The Edinburg landfill reclamation project complied with existing Occupational Health and Safety Administration (OSHA) requirements for worker health and safety. The excavated landfill did not have a waste inspection program when the landfill was filled, and a number of barrels were uncovered during the reclamation. Personal protection devices conforming to U. S. EPA Level C requirements were worn by excavation personnel. While the Edinburg demonstration project successfully reclaimed a 0.4-ha (1-acre) landfill area, a number of observations were noted for future reclamation projects. Cold weather and rain impeded operation at times. The separation efficiency for fine materials decreased during wet weather conditions. The unknown nature of the fill material required extra safety precautions that were justified by the discovery of a number of crushed drums.

Lancaster County, Pennsylvania

Pennsylvania's first landfill mining project was conducted at the Frey Farm Landfill, managed by the Lancaster County Solid Waste Management Authority.[135] The project began in 1991. By 1996 the authority had processed over 170,600 tonnes (188,600 tons) of reclaimed waste recovering 283,000 m³ (370,000 yd³) of space for reuse. The waste in the landfill was relatively young, ranging from one to five years old. The primary goal of the reclamation process was to recover waste from the landfill for use as fuel input to the local waste-to-energy (WTE) facility. The landfill functioned basically as an equalization facility for waste. When the WTE facility was operating at or above capacity, waste was landfilled as needed. When the WTE was below capacity, waste was reclaimed and burned.

Excavation was performed using a strip mining method, using a track dozer to stockpile the waste within reach of a hydraulic excavator. The excavator fed the waste into the portable trommel screen with 2.5-cm (1-in) screen openings. The recovered fine fraction from the trommel consisted primarily of cover soil since waste stabilization was minimal. This material was reused as cover soil on site. The larger material was delivered to the Resource Recovery Facility (RRF) at the WTE facility where the ferrous metals were recovered and recycled. Approximately 56 percent of recovered material was shipped to the WTE, 41 percent was recovered soil, and three percent was noncombustable materials which were required to be re-landfilled.[135]

USE OF RECLAIMED MATERIALS

As previously stated, several classes of recovered material are generated during landfill reclamation. A primary goal of the reclamation process is to safely and economically reuse the recovered materials, minimizing the amount of material that must be permanently disposed in a landfill. The oversized fraction that is collected during the screening operation contains the large waste materials which did not decompose. Some materials in this fraction may present an opportunity for recovery. Ferrous metal and aluminum have most commonly been recovered during previous landfill reclamation activities. It is important to note that the quality of the recovered materials will deteriorate during the landfilling and reclamation process, primarily as a result of contamination with cover soil. For items which may be easily separated from the rest of the waste, such as ferrous metal, recovery may prove to be an economical option. The remaining oversized materials would be disposed of in a separate unit.

The largest fraction of material will, in most cases, be the fraction composed of cover soil and degraded organic waste. It is this fraction which offers the largest potential to extend the life of the landfill site. A number of end uses are possible, dictated by available markets, the quality of the material, and the applicable regulatory standards for reuse. Available markets for the fine fraction recovered from landfill reclamation will be similar to those for other soil-like waste materials such as compost, ash, soil from remediated contamination sites, used sandblasting grit, and soil recovered from recycling of construction and demolition debris. During the process of a market evaluation, the recycling success and impediments associated with similar materials should be investigated. Since the landfill site is usually located in less populated areas with adequate site buffering, unique opportunities exist for the operator to develop reuse options on site. Examples of such options include growing sod, timber areas, and use as construction and embankment fill on site. Landfill operators who also provide other public work services may be able to create market opportunities related to these activities.

Offsite reuse may also be possible, but again, only if appropriate regulatory requirements are met. One issue of concern is the presence of impurities such as small pieces of glass, rock, or wood. While these impurities may hinder reuse in some agricultural applications, this factor may not be important if the material is used as construction fill and the proper engineering specifications are met. The quality issue which would most likely cause limitations to reuse would be the presence of trace amounts of hazardous chemicals. Most organic chemicals will be destroyed in the biologically active environment of the bioreactor, especially when exposed to the extended treatment time associated with the bioreactor landfill. Contaminants such as heavy metals, however, tend to remain in the waste if originally present in the waste stream, a common problem encountered when dealing with compost or sludge application as well.

The particular standards imposed will reflect the proposed use and the governing regulations. While specific regulations for reclaimed soil from landfills do not exist in most cases, it is expected that a number of restrictions would be imposed by the regulatory agency with jurisdiction. Regulatory limits do exist at the national level for sludge,[136] and many states have developed limits for compost, ash, and remediated soil.

■ FUTURE DIRECTIONS FOR BIOREACTOR LANDFILLS

The reclamation of bioreactor landfills is a natural extension of the concept of operating a landfill as a waste treatment system rather than a storage system. Long-term liability concerns can be minimized if waste is quickly treated to a point where further degradation will not occur or will occur so slowly that leachate contamination and gas production are no longer threats to the environment. A specified design life of 20 years for geosynthetic membranes may not provide adequate protection for the conventional landfill with stabilization periods of many decades. The potential impact on groundwater from a cleaner leachate is significantly reduced. Similarly, gas production confined to a few years rather than decades provides opportunity for control and destruction of air toxics and greenhouse gases. With sufficient data acquired through monitoring of today's bioreactor landfills, regulators may eventually reduce long-term monitoring frequency and duration for leachate recirculating landfills, recognizing the reduced potential for adverse environmental impact. Reduced liability (and associated costs required for financial assurance) and minimal monitoring will translate into significant cost savings. The reclamation of stabilized waste will effectively eliminate future liability associated with the waste, while minimizing the need to continually site, permit, and construct landfill facilities.

REFERENCES

1. Carra, J. S. and R. Cossu, Eds. *International Perspectives on Municipal Solid Wastes and Sanitary Landfilling.* Academic Press, London, 1989.
2. Pohland, F. G. *Sanitary Landfill Stabilization with Leachate Recycle and Residual Treatment.* U.S. Environmental Protection Agency, Cincinnati, EPA-600/2-75-043, 1975.
3. U.S. Environmental Protection Agency. Solid waste disposal facility criteria; proposed rule. *Fed. Reg.*, 53(168) 33313, 1988.
4. Tchobanoglous, G., H. Theisen, and S. A. Vigil, *Integrated Solid Waste Management: Engineering Principles and Management Issues.* McGraw-Hill, New York, 1993.
5. U.S. Environmental Protection Agency. *Requirements for Hazardous Waste Landfill Design, Contruction, and Closure.* Office of Research and Development, Cincinnati, EPA/625/4-89/022, 1989.
6. Bagchi, A. *Design, Construction, and Monitoring of Landfills.* John Wiley & Sons, New York, 1994.
7. McBean, E. A., F. A. Rovers, and G. J. Farquhar, *Solid Waste Landfill Engineering and Design.* Prentice Hall, Englewood Cliffs, NJ, 1995.
8. U.S. Environmental Protection Agency. *Quality Assurance and Quality Control for Waste Containment Facilities.* Office of Research and Development, Washington, DC, EPA/600-R-93/182, 1993.
9. U.S. Environmental Protection Agency. *Design and Construction of RCRA/CERCLA Final Covers.* Office of Research and Development, Washington, DC, EPA/625/4-91/025, 1991.
10. Fenn, D. K. J. Hanley, and T. V. DeGeare, *Use of the Water Balance Method for Predicting Leachate Generation from Solid Waste Disposal Sites.* U.S. Environmental Protection Agency, EPA-530-SW-169, Cincinnati, 1975.
11. Schroeder, P. R., C. M. Lloyd, and P. A. Zappi, *The Hydrologic Evaluation of Landfill Performance (HELP) Model, User's Guide for Version 3,* EPA/600/R-94/168a, 1994.
12. Hatfield, K., X Kang., W. L. Miller, and T. G. Townsend, *Hydrologic Management Models for Operating Landfills,* presented at the Florida Center for Solid and Hazardous Waste Management First Annual Solid Waste Research Symposium, Report No. 93-2, 1993.
13. Hatfield, Kirk, and Miller, W. Lamar, *Hydrologic Management Models for Operating Sanitary Landfills,* Final Report to the Florida Center for Solid and Hazardous Waste Management, November 1994.
14. Moore, C. A. *Landfill and Surface Impoundment Performance Manual.* Prepared for the U.S. EPA, Cincinnati, 1980.
15. McEnroe, B. M., P. R. Schroeder. Leachate collection in landfills: steady case. *J. of Env. Eng.* 114(5):1052-1062, 1988.
16. Pohland, F. G. and S. R. Harper. *Critical Review and Summary of Leachate and Gas Production From Landfills.* EPA/600/2-86/073, U.S. Environmental Protection Agency, Cincinnati, 1986.

17. Brown, K. W. and K. C. Donnelly. An estimation of the risk associated with the organic constituents of hazardous and municipal waste landfill leachates. *J. of Haz. Wastes and Haz. Mat.*, 5(1): 3, 1988.

18. Albaiges, J., F. Casado, and F. Ventura. Organic indicators of groundwater pollution by a sanitary landfill. *Water Res.* 20(9): 1153, 1986.

19. Schultz, B. and P. Kjeldsen. Screening of organic matter in leachate from sanitary landfills using gas chromatography combined with mass spectrometry. *Water Res.* 20(8): 965, 1986.

20. Harmsen, J. Identification of organic compounds in leachate from a waste tip. *Env. Sci. and Tech.*, 17(6): 699, 1983.

21. Sawney, B. L. and R. P. Kozloski. Organic pollutants in leachates from landfill sites. *J. of Env. Qua.* 13(3): 349, 1984.

22. Oman, Cecilia and P. Hynning. Identification of organic compounds in municipal landfill leachates. *Env. Pollution* 80: 265-271, 1993.

23. Chian, E. S. K. Stability of organic matter in landfill leachates. *Water Res.*, 11(2): 159, 1977.

24. Lu, J. C. S., B. Eichenberger, and R. J. Stearns. *Leachate from Municipal Landfills: Production and Management, Pollution Technology Review No. 119,* Calscience Research — Noyes Publications, New Jersey, 1985.

25. Bennett, G. F. Air quality aspects of hazardous waste landfills. *J. of Haz. Waste and Haz. Mat.*, 4(2): 119, 1987.

26. Young P. J. and A. Parker. The identification and possible environmental impact of trace gases and vapors in landfill gas. *Waste Mgt. and Res.*, 1: 213, 1983.

27. Wood, J. A. and M. L. Porter. Hazardous pollutants in Class II landfills. *J. of the Air Pollution Cntrl. Assn..* 37(5): 609, 1987.

28. LaRegina, J. and J. W. Bozzelli. Volatile organic compounds at hazardous waste sites and a sanitary landfill in New Jersey. *Env. Progress,* 5(1): 18, 1986.

29. Siu, W, D. A. Levaggi, and T. F. Brennan. *Solid waste assessment test results from landfills in the San Francisco Bay area.* Presented at the 82nd Annual meeting and exhibition of the Air and Waste Management Association, Anaheim, CA, 1989.

30. Mata-Alvarez J. and A. Martina-Verdure. Laboratory simulation of municipal solid waste fermentation with leachate recycle. *J. Chem. Tech. Biotechnol.*, 36: 547-556, 1986.

31. Barlaz, M. A., R. K. Ham, and D. M. Schaefer. Methane production from municipal refuse: a review of enhancement techniques and microbial dynamics. *Crit. Rev. in Env. Cntrl.*, 19(6): 557, 1990.

32. Buivid, M. G., D. L. Wise, M. J. Blanchet, E. C. Remedios, B. M. Jenkins, W. F. Boyd, and J. G. Pacey. Fuel gas enhancement by controlled landfilling of municipal solid waste. *Res. & Cons.*, 6: 3, 1981.

33. Chian, E. S. and F. B. DeWalle. Characterization of soluble organic matter in leachate. *Env. Sci. and Tech.*, 11(2): 159, 1977.

34. Klink, R. E. and R. K. Ham. Effect of moisture movement on methane production in solid waste landfill samples. *Res. & Cons.* 8: 29, 1982.

35. Tittlebaum, M. E. Organic carbon content stabilization through landfill leachate recirculation. *J. of Water Poll. Cntl. Fed.* 54: 428, 1982.

36. Doedens, H. and K. Cord-Landwehr. Leachate recirculation. In *Sanitary Landfilling: Process, Technology and Environmental Impact*, T. H. Christensen, R. Cossu, and R. Stegmann, Eds., Academic Press, London, 1989.

37. Otieno, F. O. Leachate recirculation in landfills as a management technique. In *Proceedings of Sardinia '89, Second International Landfill Symposium*, Calgari, Italy, 1989.

38. Pohland, F. G., W. H. Cross, J. P. Gould, and D. R. Reinhart. *The Behavior and Assimilation of Organic Priority Pollutants Codisposed with Municipal Refuse*. U.S. EPA, EPA Coop. Agreement CR-812158, Volume 1, 1992.

39. IEA Expert Working Group on Landfill Gas. *Data Base of Landfill Test Cells*.1995.

40. EMCON. *Twelve-month extension Sonoma County solid waste stabilization study* (final report). GO6-EC-00351, San Jose, CA, 1976.

41. Leckie, J. O., J. G. Pacey, and C. Halvadakis. Landfill management with moisture control. *J. of Env. Eng.* 105(EE2): 337, 1979.

42. Pohland, F. G. Leachate recycle as landfill management option. *J. of Env. Eng.* 106(EE6):1057-1069, 1980.

43. EMCON. *Controlled landfill project — fifth annual report.* Project 343-03.02, San Jose, CA, 1987.

44. Pacey, J. G., J. C. Glaub, and R. E. Van Heuit. Results of the Mountain View controlled landfill project. In *Proceedings of the GRCDA 10th International Landfill Gas Symposium*, GRCDA, Silver Spring, MD, 1987.

45. New York Energy Research and Development Authority. *Enhancement of Landfill Gas Production, Nanticoke Landfill, Binghamton, NY.* NYSERDA Report 87-19, Wehran Engineering, July 1987.

46. Lechner, P., T. Lahner, and E. Binner. *Reactor Landfill experiences gained at the Breitnau Research Landfill in Austria.* Presented at the 16th International Madison Waste Conference, Madison, WI, 1993.

47. Campbell, D. J. V. UK Brogborough test cell project. In *Landfill Gas Enhancement Test Cell Data Exchange, Final Report of the Landfill Gas Expert Working Group.* P. Lawson, Ed., International Energy Agency, Harwell Laboratory, Oxfordshire, U.K., 1991.

48. Brundin, H. The SORAB test cells. In *Landfill Gas Enhancement Test Cell Data Exchange, Final Report of the Landfill Gas Expert Working Group,* P. Lawson, Ed., Oxfordshire, U.K.: International Energy Agency, Harwell Laboratory, 1991.

49. Natale, B. R. and W. C. Anderson. *Evaluation of a Landfill with Leachate Recycle*. Draft report to U.S. EPA Office of Solid Waste, 1985.

50. Barber, C., and P. J. Maris. Recirculation of leachate as a landfill management option: benefits and operational problems. *Q. J. Eng. Geol., London*, 17:19-29, 1984.

51. Watson, R. *State of Delaware's case history review: a full scale active waste management approach.* Presented at the Modern Double Lined Landfill Management Seminar, Saratoga Springs, NY, 1993.

52. Stegmann, R. and H. H. Spendlin. Enhancement of degradation: German experiences. In *Sanitary Landfilling: Process, Technology and Environmental Impact.* T. H. Christensen, R. Cossu, and R. Stegmann, Eds., Academic Press, London, 1989.

53. Miller, W. L., T. Townsend, J. Earle, H. Lee, D. R. Reinhart, and P. Paladugu. *Leachate recycle and the augmentation of biological decomposition at municipal solid waste landfills.* Presented at the First Annual Research Symposium, Florida Center for Solid and Hazardous Waste Management, Orlando, 1993.

54. Townsend, T. G., W. L. Miller, J. F. K. Earle. Leachate Recycle Infiltration Ponds, *J. of Env. Eng.,* 121(6), 465-471, 1995.

55. Townsend, T. G. *Leachate Recycle at the Southwest Landfill Using Horizontal Injection.* PhD Thesis, University of Florida, Gainesville, 1995.

56. Townsend, T. G., W. L. Miller, R. A. Bishop, J. H. Carter. Leachate Recycle Infiltration Ponds, *Solid Waste Tech.,* 18-24, July 1994.

57. Lagerkvist, A. Test cells in Sweden, status report, October 1991. In *Landfill Gas Enhancement Test Cell Data Exchange, Final Report of the Landfill Gas Expert Working Group,* P. Lawson, Ed., Oxfordshire, U.K.: International Energy Agency, Harwell Laboratory, 1991.

58. Lagerkvist, A. Two step degradation — an alternative management technique. In *Proceedings Sardinia '91, Third International Landfill Symposium,* Calgari, Italy, 1991.

59. Willumsen, H. C. Registration and optimizing of gas production from closed waste test cells in landfills. In *Landfill Gas Enhancement Test Cell Data Exchange, Final Report of the Landfill Gas Expert Working Group,* P. Lawson, Ed., Oxfordshire, U.K.: International Energy Agency, Harwell Laboratory, 1991.

60. Cossu, R. and G. Urbini. Sanitary landfilling in Italy. In *International Perspectives on Municipal Solid Wastes and Sanitary Landfilling,* Carra, J. S. and R. Cossu, Eds,. Academic Press, London, 1989.

61. Fergusen, R. G. Sanitary landfilling in Canada. In *International Perspectives on Municipal Solid Wastes and Sanitary Landfilling,* Carra, J. S. and R. Cossu, ed,. Academic Press, London, 1989.

62. Gonzales, A. J. *An integrated system for leachate reintroduction and gas venting.* Presented at the 1994 GEOENVIRONMENT 2000 Conference, New Orleans, LA, USA, 1994.

63. Kmet, P., *EPA's 1975 Water Balance Method: Its Use and Limitations.* A Wisconsin Department of Natural Resources Guidance Report, Madison, October, 1982.

64. Quasim, S. R. and J. C. Buchinal, Leaching from Simulated Landfills, *J. Water Poll. Cntl. Fed.,* 42(3): 371, 1970.

65. Reinhardt, J. J. and R. K. Ham. *Solid Waste Milling and Disposal on Land without Cover.* U.S. Environmental Protection Agency, Cincinnati, PB-234 930, 1974.

66. Holmes, R., The Absorptive Capacity of Domestic Refuse from a Full-Scale, Active Landfill, *Waste Mgt.,* 73(11): 581, 1983.

67. Korfiatis, G. P. , A. C. Demetracopolous, E. L. Bourodimos, and E. G. Nawey. Moisture Transport in a Solid Waste Column. *J. of Env. Eng.,* 110(EE4):789-796, 1984.

68. Oweis, E. S., D. A. Smith, R. B. Ellwood, and D. S. Greene. Hydraulic characteristic of municipal refuse. *ASCE J. Geotechnical Eng.,* 116(4): 539-553, 1990.

69. Walsh, J. J. and R. N. Kinman, Leachate and Gas Production under Controlled Moisture Conditions, *in Proceedings of the Fifth Annual Research Symposium Municipal Solid Waste Land Disposal,* EPA-600/9-79-023a, 1979.

70. Remson, I., A. A. Fungaroli, and A. W. Lawrence, Water Movement in an Unsaturated Sanitary Landfill, *J. of the Sanitary Eng. Div., ASCE* 94(SA2): 307, 1968.

71. Canziani, R. and R. Cossu, Landfill Hydrology and Leachate Production. In *Sanitary Landfilling: Process, Technology and Environmental Impact,* T. H. Christensen, R. Cossu, and R. Stegmann, Eds., Academic Press, London, 1989.

72. Fungaroli, A. A. and R. L. Steiner. *Investigation of sanitary landfill behavior.* EPA/600/2-79, 1979.

73. Wigh, R. J. and D. R. Brunner, Leachate Production from Landfilled Municipal Waste — Boone County Field Site, *in Proceedings of the Fifth Annual Research Symposium Municipal Solid Waste Land Disposal,* EPA-600/9-79-023a, 1979.

74. Hentrich, R. L., J. T. Swartzbaugh, and J. A. Thomas, Influence of MSW Processing on Gas and Leachate Production, in *Proceedings of the Fifth Annual Research Symposium Municipal Solid Waste Land Disposal,* EPA-600/9-79-023a, 1979.

75. Straub, W. A. and Lynch, D. R., *Models of Landfill Leaching: Moisture flow and Inorganic Strength,* Environmental Engineering Div., Vol. 108, pp. 231-250, 1982.

76. CMA Engineers, *Leachate Recirculation Annual Report.* Submitted to NHDES Waste Management Division, Portsmouth, NH, 1993.

77. Noble, J. J. and A. E. Arnold. Experimental and mathematical modeling of moisture transport in landfills. *Chemical Eng. Comm.* 100:95-111, 1991.

78. Al-Yousfi, A. B. *Modeling of leachate and gas production and composition at sanitary landfills,* Pittsburgh, PA: University of Pittsburgh, PhD Thesis, 1992.

79. Oweis, I. and Khera R. Criteria for geotechnical construction of sanitary landfills. in *Int. Symp. on Envir. Geotech.,* H. Y. Fang, Ed., Lehigh Univ. Press, Bethlehem, PA, 1986, 205-222.

80. Wehran Engineering, P. C. & Dynatech Scientific, Inc. *Enhancement of Landfill Gas Production, Nanticoke Landfill, Binghamton, NY.* New York State Energy Research and Development Authority, NYSERDA Report 87-19. July 1987.

81. Bleiker, D. E., E. McBean, and G. Farquhar. Refuse sampling and permeability Testing at the Brock West and Keele Valley Landfills. presented at the Sixteenth International Madison Waste Conference, Madison, WI, September 22-23, 1993.

82. Voss, Clifford I., *SUTRA, Saturated-Unsaturated Transport, A Finite-Element Simulation Model for Saturated-Unsaturated, Fluid-Density-Dependent Ground-Water Flow with Energy Transport or Chemically Reactive Single-Species Solute Transport,* U.S. Geological Survey, National Center, Reston, Va., 1984.

83. Voss, Clifford I., *SUTRA-SUTRA Plot — IGWMC - FOS 27 EXT,* International Ground Water Modeling Center, 1991.

84. Lappala, E. G., Healy, R. W., and Weeks, E. P, *Documentation of Computer Program VS2D to Solve the Equations of Fluid Flow in Variably Saturated Porous Media*, Water-Resources Investigations Report 83-4099, U.S. Geological Survey, 1987.

85. Miller, W. L., T. Townsend, J. Earle, H. Lee, and D. R. Reinhart. *Leachate recycle and the augmentation of biological decomposition at municipal solid waste landfills*. Presented at the Second Annual Research Symposium, Florida Center for Solid and Hazardous Waste Management, Tampa, 1994.

86. King, D. and A. Mureebe. Leachate management successfully implemented at landfill. *Water Env. and Tech.* 4(9):42, 1992.

87. Gould, J. P., W. H. Cross, and F. G. Pohland. Factors influencing mobility of toxic metals in landfills operated with leachate recycle. In *Emerging Technologies in Hazardous Waste Management*, D. W. Tedder and F. G. Pohland, Ed., ACS Symposium Series 422, 1989.

88. Pohland, F. G., S. R. Harper, K. Chang, J. T. Dertien, and E. S. K. Chian. Leachate generation and control at landfill disposal sites. *Water Pollution Res. J. of Canada* 20(3): 10-24, 1985.

89. Reinhart, D. R. *Active Municipal Waste Landfill Operation: A Biochemical Reactor*. U.S. Environmental Protection Agency, Cincinnati, 1996.

90. Sulfita, J., C. Gerba, R. Ham, A. Palmisano, W. Rathje, and J. Robinson. The world's largest landfill. *Env. Sci. & Tech.* 26(8):1486-1495, 1992.

91. Lema, J. M., Mendez, R., and Blazquez, R. Characteristics of landfill leachates and alternatives for their treatment: a review. *Water, Air and Soil Pollution*, 40:223-250(1988).

92. Venkataramani, E. S., Ahlert, R. C., and Corbo, P. Biological treatment of landfill leachates. *CRC Crit. Rev. in Env. Cntrl.*, 14(4):333-372, 1986.

93. Maier, T. Expected Benefits of a Full-Scale Bioreactor Landfill. Presented at SWANA's 34th Annual International Solid Waste Exposition, Portland, OR, 1996.

94. Palumbo, D. *Estimating Early MSW Landfill Gas Production*. MS Thesis, University of Central Florida, Orlando, 1995.

95. Koerner, G. R. and R. M. Koerner. Permeability of granular drainage material. Presented at the U.S. EPA Bioreactor Landfill Design and Operation Seminar, Wilmington Delaware, 1995.

96. Giroud, L. P. "Granular filters and geotextile filters," *Proceedings of GeoFilters '96*. Lafleur, J. and Rollin, A. L., Editors, Montreal, PQ, Canada, 1996, pp 565-580.

97. Giroud, J. P., A. Khatami, and K. Badu-Tweneboah. Evaluation of the rate of leakage through composite liners. *Geotextiles and Geomembranes*, 8(4): 337-340, 1989.

98. Baetz, B. W. and K. A. Onysko. Storage volume sizing for landfill leachate-recirculation systems. *J. of Env. Eng.* 119 (2): 378-383, 1993.

99. Robinson, H. D. and P. J. Maris. The treatment of leachates from domestic waste in landfill sites. *J. of Water Poll. Cntl. Fed.* 57(1): 30, 1985.

100. Miller, D. and Emge. *Leachate Recirculation System Design, Operation and Performance at the Kootenai County (ID) Landfill*. Presented at SWANA's 34th Annual International Solid Waste Exposition, Portland, OR, 1996.

101. Miller, L. V., R. E Mackey, and J. Flynt. *Evaluation of a PVC liner and leachate collection system in a 10 year old municipal solid waste landfill*. Presented at the 29th Annual Solid Waste Assn of North America Solid Waste Exposition, 1991.

102. Leszkiewicz, J. and P. McAulay. *Municipal solid waste landfill bioreactor technology closure and post-closure*. Presented at the U.S. EPA Seminar — Landfill Bioreactor Design and Operation, Wilmington, DE, 1995.

103. Brown, K. S. And W. E. Clister, *Design Procedures for Landfill Gas Interception, Collection, and Extraction Systems Pursuant to the Proposed EPA Regulations for the Control of Landfill Gas*. Presented at the SWANA 16th Annual landfill Gas Symposium, Louisville, K, 1993.

104. Kraemer, T. A. *Gas Collection Beneath a Geomembrane Final Cover System*. Presented at the SWANA 16th Annual landfill Gas Symposium, Louisville, K, 1993.

105. Lewis, K. A. *The effects of landfill management strategies on landfill gas utilization alternatives*. M.S. Thesis, University of Central Florida, Orlando, 1995.

106. Augenstein, D., J. Pacey, R. Moore, and S. Thorneloe. Landfill methane enhancement. In *Proceedings from the SWANA 16th Annual Landfill Gas Symposium*, GR-LG 0016, Louisville, KY, 1993, 21-48.

107. Baetz, B. W. and P. H. Byer. Moisture control during landfill operation. *Waste Mgt. & Res.* 7: 259-275, 1989.

108. Rattray, T. Demographics and discards. *Garbage* 4(6): 27, 1993.

109. Franklin Associates, Ltd. *Characterization of municipal solid waste in the United States — 1994 Update*. Prepared for the U.S. Environmental Protection Agency EPA530/S-94/042, 1994.

110. U.S. Congress, Office of Technology Assessment. *Facing America's Trash: What Next for Municipal Solid Waste?* Office of Solid Waste and Emergency Response, Washington, DC, PB90-174897, 1990.

111. DeWalle, F. B., E. S. K. Chian, and E. Hammerberg. Gas production from solid waste in landfills. *J. of Env. Eng.*, 104(EE3):415-432, 1978.

112. Bookter, T. J. and R. K. Ham. Stabilization of solid waste in landfills, *J. of Env. Eng.*, 108(6) 1089, 1982.

113. Merz, R. C. and R. Stone. *Special Studies of a Sanitary Landfill*, U.S. Department of Health, Education, and Welfare, Washington, DC, 1970.

114. Fletcher, P. Landfill gas enhancement technology — laboratory studies and field research. In *Proceedings of Energy from Biomass and Wastes*, XIII, IGT, 1989. pp. 1001.

115. American Technologies, Inc. *Aerobic Landfill Bioreactor Pilot Project, Baker Place Road Landfill — Columbia County, Georgia*. Internet address: http://www.atechinc.com, 1996.

116. Rovers, F. A. and G. J. Farquhar. Infiltration and landfill behavior. *J. of Env. Eng.* 99(10): 671-690, 1973.

117. Rees, J. F. Optimisation of methane production and refuse decomposition in landfills by temperature control. *J. Chem. Tech. Biotechnol.* 30: 458-465, 1980.

118. Doedens, H. and K. Cord-Landwehr. Leachate recirculation. In *Proceedings Sardinia '91, ISWA International Sanitary Landfill Symposium*, Calgari, Italy, 1987.

119. Pohland, F. G. *Landfill Bioreactors: Historical Perspective, Fundamental Principles, and New Horizons in Design and Operation*. Presented at the U.S. EPA Seminar — Landfill Bioreactor Design and Operation, Wilmington, DE, 1995.

120. Agora, Santosh. *Mathematical modeling of leachate recirculating landfills*. M.S. Thesis, University of Central Florida, Orlando, 1995.

121. Hartz, K. E., Klink, R. E., and R. K. Ham. Temperature effects: methane generation From landfill samples. *J. of Env. Eng.* 108(EE4):629-638, 1982.
122. Gurijala, K. Rao. and J. M. Suflita. Environmental factors influencing methanogenesis from refuse in landfill samples. *Env. Sci. & Tech.,* 27(6):1176-1181, 1993.
123. Pohland, F. G. *Assessment of Solid Waste and Remaining Stabilization Potential after Exposure to Leachate Recirculation at a Municipal Landfill.* Prepared for Post, Buckley, Schuh & Jernigan, Inc, Project No. 07-584.18, 1992.
124. Leuschner, A. P. Enhancement of degradation: laboratory scale experiments, In *Sanitary Landfilling: Process, Technology and Environmental Impact.* T. H. Christensen, R. Cossu, and R. Stegmann, Eds., Academic Press, London, 1989.
125. ten Brummeler E. T., C. J. Horbach, and I. W. Koster. Dry anaerobic batch digestion of the organic fraction of municipal solid waste. *J. of Chem. Tech. and Biotechnology* 50:191-209, 1991.
126. Dwyer, J. R., J. C. Walton, W. E. Greenberg, and R. Clar., *Evaluation of Municipal Solid Wate Landfill Cover Designs,* U.S. EPA Hazardous Waste Engineering Research Laboratory, Cincinnati, 1986.
127. Augenstein, D. and R. Yazdani. Landfill Bioreactor Instrumentation and Monitoring. Presented at the U.S. EPA Seminar — Landfill Bioreactor Design and Operation, Wilmington, DC, 1995.
128. Owens, J. M. and D. P. Chynoweth. Biochemical methane potential of MSW components. Presented at the International Symposium on Anaerobic Digestion of Solid Waste, Venice, Ital, April 15–17, 1992.
129. Nelson, H. Landfill Reclamation Strategies. *BioCycle,* October 41-44, 1994.
130. Guerriero, J. R., *Landfill Reclamation and Its Applicability to Solid Waste Management.* Malcolm Pirnie, 1994.
131. Fahey, R. E., Case Study: Collier County's Landfill Reclamation Project. Presented at the National Conference on Reclaiming Landfills, Chicago, March 1993.
132. U.S. Environmental Protection Agency, *Evaluation of the Collier County, Florida, Landfill Mining Demonstration.* EPA/600/R-93/163, Sept., 1993.
133. Stessel, R. I. and Murphy, R. J., *Processing Material Mined from Landfills.* Interim Report Volume II. University of South Florida, Tampa, Sept. 1991.
134. New York State Energy Research and Development Authority (Energy Authority). *Town of Edinburg Landfill Reclamation Report.* Energy Authority Report. May 1992.
135. Foster, G. A., et. al. *Assessment of Landfill Mining and the Effects of Age on the Combustion of Recovered Municipal Solid Waste.* Lancaster County Solid Waste Management Authority. Lancaster County, PA., 1994
136. U.S. Environmental Protection Agency. Standards for the Use or Disposal of Sewage Sludge. *Fed. Reg.,* 58(32), p.9248, 1993.

INDEX

A

Absorptive capacity of waste, leachate generation, 85
Acclimation period, landfill stabilization phases, 16
Acid generation/acids
  experimental systems, *see* specific systems
  landfill stabilization phases, 16–17, 19
  leachate characteristics, *see* Leachate, characteristics of recirculating landfills
  operations
    monitoring, 153
    recirculation strategies, 141–143
    redox conditions and, 139–140
    waste placement rate and, 143
Activated carbon, 20
Adjustment phase, landfill stabilization phases, 16
Aerobic conditions
  landfill stabilization phases, 16, 19
  operational considerations, 139–140
Age of landfill
  and leachate characteristics, 18
  shifts in leachate composition, 19
Air injection, 139
Alachua County, Florida, 61–64, 105, 113, 126–127, 132
Al Turi Landfill, Orange County, New York, 113
Anaerobic conditions
  gas collection system design, 134
  landfill stabilization phases, 16, 19
  leachate treatment, 113
  operational considerations, 139–140
Animal vectors of disease, 2
Artesian conditions, horizontal trench design, 129
Ash, 1
Augmentation of bioreactor, 148–149
Austria, 40, 43–44, 59, 105, 110

B

Baker Place Road landfill, Columbia County, Georgia, 139
Barriers
  cap design, 131
  compacted soil, 10
  containment systems, 9–11
  features of landfill, 9, 135–136
  synthetic, 10–11
Binghamton, New York, 39–4043, 59
Biofouling, 127
Biological decomposition, *see also* Degradation of waste
  cap design, 131
  leachate generation, 85
Biological methane potential (BMP), 115, 152
Biological oxygen demand (BOD)
  experimental systems, *see* specific systems
  monitoring, 153
  stabilization of landfill, 19
Biological system, landfill as, 16–18
Biomass, landfill stabilization phases, 16
Bioreactor, 1
  augmentation of, 148–149
  design of, *see* Design of bioreactor
  landfills as, 3–4
Blow down, 2
BOD, *see* Biological oxygen demand
Bornhausen Landfill, Germany, 56–57, 60
Breitenau Landfill, Austria, 40, 43–44, 59, 105
  cover, 150
  waste stabilization rates, 110
Broghborough, United Kingdom, 43, 45, 59
Buffering, 2
  and gas generation, 20
  operations, bioreactor augmentation, 149

177

C

M

N

Y